STELIANA TASE

Mathematics guide and summary book

First Part

ROMANIAN TEXTBOOKS
"HOLY MARTYRS BRÂNCOVENI" SCHOOL

Constanța
2015

The Orthodox Association of Parents and Teachers „St. Grigorie Palama"
e-mail: - asociatie@manualeromanesti.ro
        - manualeromanesti@gmail.com
Phone: +40.731.749.328
www.manualeromanesti.ro
ISBN-13: 978-1517731779
ISBN-10: 1517731771

PRESENTATION

Usually the preface of a book is not read. Most people rush to get to the book content. If you are reading this, then you do not fit into the category aforementioned. It means that you do not rush to the exercises part, but you first want to understand the author's intention.

This book is a special work. It is neither a textbook, nor a collection of exercises, but a *guide and summary book.* What does it mean to guide? *To guide* is to show the right way and give help on the way. Mathematics is a very beautiful science, but, as Academician Solomon Marcus well says, there is in it a beauty hidden for students, due to wrongly put together school programs, overloaded with irrelevant and non-pragmatic issues.

The book you are reading shows you an elegant way to the math in middle school, so that you enjoy the beauty of this science. When I first read the manuscript, I remembered the special lessons of the 90s, the great educators, professors at the Faculty of Mathematics in Bucharest. I am talking about Prof. Dr. Laurentiu Panaitopol and Assoc. Prof. Dr. Gheorghe Mocanu.

In the pages of this guide you are reading, I found the elegance, concision and clarity of the lessons of these great teachers. Taught in this way, mathematics can be easily memorized, therefore *Guided Book* is a good *memorizing tool* as well.

I recommend this book to mathematics teachers, parents and students, hopping that it would be very useful for everyone. For teachers, it is a working instrument, for parents it is a supporting tool and for students, a mentor and memory book.

Mathematics should not be a cause for animosity and envy because of a wild competition, led by the law of the jungle ("the strongest wins"). Chasing awards for awards' sake only is not specific to the genuine intellectual environment.

First of all, Mathematics is a method of disciplining the mind, and its results (applied physics, chemistry, biology, medicine, pedagogy, psychology, economics) should only be used for the good of humanity.

These applications require teamwork and more students should be advised to cooperate than to compete, thus making science contribute to the strengthening of ties between people.

Prof. Ioan Vlăducă,
Scientific Director of
School "Holy Martyrs Brâncoveni"

FOREWORD

Why did we title this *Mathematics Guide and Memory book*? Theory will be followed by exercises and examples so that students will understand how to apply it and I hope they will not work mechanically. If students learn theory very well, the ideas will come as soon as they read the hypothesis.

For example: if we have a division with remainder, we will think of Euclidean division; if we have bisectrices and parallels, we'll think about equal angles and the angles made by parallels intersecting a secant.

Combining theory with exercises and examples, this *Mathematics Guide and Summary book* could be used by all students, including the ones less talented in math: if they learn very well the theory, then they will realize that math is not difficult! Go ahead and try and you will see that I am right, all the more so as there is not very much theory in the book, some formulas we deduce and some will be easier to remember.

Before I started working on this memory book, I received the blessing of our kind and remarkable Archimandrite Arsenie Papacioc. Here is what he said:

> "The root of all evils is ignorance. It is not the love of money, but ignorance, say the Holy Fathers. Because if you did not know, you should motivate why you did not know: was it because you were lazy, you did not study, guided by those who had that position and means to say something to you. So laziness will cost you, it is a great sin, as we say so: "the holy" laziness and "the Great Martyr sleep " ... And then, of course, we do our duty and everyone is well, we learn more math because it is a positive science, no deviations are allowed, "I'll see..." doesn't work, you must be informed, you have to know. It's good to know math, because everywhere you go you have to make calculations, everywhere you need to make your own reckonings, especially in your private problems..."

And I say that there is no student who knows mathematics and have low grades at the other subjects, because math is a sport of the mind and helps you think about all the other matters and always in life, at any time. I also say that if the Good Lord has given us our mind, it is a pity not to use it.

I wish you are successful in your quest!

Professor Steliana Tase

MATHEMATICS GUIDE AND SUMMARY BOOK

Contents

LOGICAL SYMBOLS AND MATHEMATICS

Symbol	Meaning	Examples
$\Rightarrow$	Implies	$x - 1 = 2 \Rightarrow x = 3$
$\Leftrightarrow$	Equivalent	$x = 2 \Leftrightarrow x + 1 = 3$
$\forall$	For all	$\forall n$ for whatever n
$\exists$	There exists	$\exists n$ exists n
$\nexists$	There does not exists	$\nexists n$ doesn't exist n
$=$	Equal	$2 = 2$
$\neq$	Not equal	$2 \neq 3,\ 2 \neq 1$
$+$	Addition	$2 + 3 = 5$
$-$	Subtraction; the oposyte of a number	$4 - 1 = 3$; -2 is the oposyte of 2
$\cdot$	Multiplication	$2 \cdot 3 = 6$
$:$	Division	$6 : 3 = 2$
$<$	Less than	$2 < 3$
$>$	Greater than	$5 > 2$
$\leq$ sau $\leqq$	Less than or equal to	$2 \leq 3;\ 2 \leq 2$
$\geq$ sau $\geqq$	Greater than or equal to	$4 \geq 1$ $4 \geq 4$
$\square$	Divisible	$12 \square 3$
$\not/$	Not divisible	$7 \not/ 2$
$\mid$	Divide	$3 \mid 12$
$\nmid$	Not divide	$2 \nmid 7$
$\mid$	Performing an operation to both members of equality	$x + 2 = 5$ |-2 $x = 3$
(m, n)	The greatest common divisor of numbers m, n. (m, n) = GCD (m, n)	$(4, 6) = 2$
$[m, n]$	The least common multiple of numbers m, n. $[m, n]$ = LCM (m, n)	$[2, 5] = 10$

{x_1, ..., x_n}	The set composed of elements x1, x2, ..., xn	A = {1, 3, 7}
$\in$	Set membership. $x \in M$, x membership of set M	$2 \in \{3, 2, 7\}$
$\notin$	No set membership	$5 \notin \{1, 3\}$
$\subseteq$	Subset: subset has fewer elements or equal to the set $A \subseteq B$ A is subset of B	$\{1,2\} \subseteq \{1,2\}$
$\subset$	Proper subset / strict subset: subset has fewer elements than the set	$\{1\} \subset \{1,2\}$
$\not\subseteq$	Not subset: left set not a subset of right set. $A \not\subseteq B$ A not a subset of B	$\{1,2\} \not\subseteq \{2, 3\}$
$\cup$	Union. A∪B: objects that belong to set A or set B	$\{1, 2\} \cup \{2, 3\} = \{1, 2, 3\}$
$\cap$	Intersection. A∩B: objects that belong to set A and set B	$\{1, 2\} \cap \{2, 3\} = \{2\}$
\	Relative complement. A\B: objects that belong to A and not to B	$\{1, 2, 3\} \setminus \{2\} = \{1, 3\}$
$C_B A$	Complementary set A to B, for $A \subset B$	A = {1}, B= {1, 2, 3} $C_B A$ = {2, 3}
A X B	Cartesian product of set A and B: set of all ordered pairs from A and B	$\{1\} \times \{2, 4\} = \{(1, 2), (1, 4)\}$
AΔB	Symmetric difference: objects that belong to A or B but not to their intersection	$\{1, 2\} \Delta \{2, 4\} = \{1, 4\}$
Φ	Empty set: Ø = { }	$\{1, 2\} \cap \{3, 4\} = \Phi$

Card M or $|M|$	Cardinality: the number of elements of set M	Card $\{4, 5, 7\} = 3$
P (A)	Power set. All subsets of A	P $(\{1, 2\}) = \{\Phi, \{1\}, \{2\}, \{1, 2\}\}$
a^n	Power $a^n = a \cdot a \cdot \ldots \cdot a$ (n factors)	$2^3 = 2 \cdot 2 \cdot 2 = 8$
$\dfrac{a}{b}$ or a/b	Fraction	$1/2 = 0, 5$
Δ	Triangle	Δ ABC
$\perp$	Perpendicular lines (90º angle)	$a \perp b$
$\parallel$	Parallel lines	$a \parallel b$

Chapter I. THE SET OF NATURAL NUMBERS

1.1. The natural numbers

The **natural** numbers are : {0, 1, 2, 3, …, 9, 10, 11, … }.

These can represent real objects or elements from nature.[1] The symbol of this set is **N**, and **N*** means the set of non-zero natural numbers, which does not contain zero. The digits are : {0, 1, 2, 3, 4, 5, 6, 7, 8, 9} .

Pay attention! Some students do not make the distinction between digits and numbers. The number 12 has two digits: the ones digit (2) and the tens digit (1).

The **even** numbers are: 0, 2, 4, 6, 8, 10, 12, …

The **odd** numbers are: 1, 3, 5, 7, 9, 11, 13, 15, …

An unknown even number is denoted by 2k, where k is a natural number. For example, for k = 3, we get the even number $2k = 2 \cdot 3 = 6$. The even numbers can be written as: {0, 2, 4, … , 2k, 2k + 2,… }

The unknown odd numbers are denoted by 2k + 1 or 2k − 1, where k is a natural number. The odd numbers can be written as: {1, 3, 5, … , 2k + 1, 2k + 3, … }.

For example, for k = 2, $2k + 1 = 2 \cdot 2 + 1 = 4 + 1 = 5$, and $2k - 1 = 2 \cdot 2 - 1 = 4 - 1 = 3$.

The **consecutive** numbers are denoted by: a, a + 1, a + 2, a + 3,… ; the odd or consecutive even numbers are denoted as: a, a + 2, a + 4, a + 6,… .

Example:

5, 6, 7, 8 are four consecutive natural numbers;

4, 6, 8 are three even consecutive numbers;

3, 5, 7 are three odd consecutive numbers.

---

[1] For example 5 apples, 4 cherries, 10 children. Negative whole numbers also exist: -1, -2, -3, etc., which can represent temperatures. For example -10° C.

Solved exercises

1) Find four natural consecutive numbers, knowing that their sum is 94.

$a + (a+1) + (a+2) + (a+3) = 94 \Rightarrow a+a+1+a+2+a+3 = 94 \Rightarrow 4a + 6 = 94 \Rightarrow 4a = 94 - 6$ (let's not forget that when we transpose elements of the equation to the left or right hand side, we change signs) $4a = 88 \Rightarrow a = 88 : 4 \Rightarrow a = 22$. Hence, the numbers are: 22, 23, 24, 25.

2) Find three even consecutive numbers, whose sum is 138.

$a + (a + 2) + (a + 4) = 138 \Rightarrow 3a + 6 = 138 \Rightarrow 3a = 138 - 6 \Rightarrow 3a = 132 \Rightarrow a = 132 : 3, \quad a = 44$. Hence, the numbers are: 44, 46, 48 .

3) Find four odd consecutive numbers, knowing that their sum is 80.

$a + (a + 2) + (a + 4) + (a + 6) = 80 \Rightarrow 4a + 12 = 80 \Rightarrow 4a = 80 - 12 \Rightarrow 4a = 68, \quad a = 17 \Rightarrow$ the numbers are 17, 19, 21, 23 .

1.2 The decomposition of a natural number in base 10

An unknown two-digit number, written in base 10, is denoted $\overline{ab}$ and is decomposed as follows: $\overline{ab} = 10a + b$. If the number is a three-digit number, we will have: $\overline{abc} = 100a + 10b + c$. Now we will explain why it decomposes this way: because **a** fills up the place of the hundreds, **b** fills up the place of the tens and **c** fills up the place of the units. If the number is a four-digit number, we will have: $\overline{abcd} = 1000a + 100b + 10c + d$

Exercises

$\overline{ab} + 2\overline{ab} + 3\overline{ba} = 330$. Find the number $\overline{ab}$.

$10a + b + 2(10a+b) + 3(10b + a) = 330$

$10a+b + 20a +2b + 30b + 3a = 330$

$33a + 33b = 330 \Rightarrow 33(a+b) = 330 \mid : 33 \Rightarrow a+b = 10$, because a and b are digits, we have the following solutions:

if $a=1 \Rightarrow b = 9 \Rightarrow \overline{ab} = 19$

$a=2 \Rightarrow b = 8 \Rightarrow \overline{ab} = 28$, analog for the others.

We obtain the solutions:

$$\overline{ab} = \{19, 28, 37, 46, 55, 64, 73, 82, 91\}$$

1.3 Operations with natural numbers

1.3.1 Addition and subtraction

The 'like' terms are :
- the terms which have the same unknown (letter);
- free terms (the natural numbers which have no letters).

Like terms are reduced. Ex: $2a + 3b + 4a - b + 9 + 5b - 3 = 6a + 7b + 6$, the like terms have been reduced. Out of 2 terms which contained the unknown a, we obtained one, hence from $2a + 4a = 6a$; out of 3 like terms, the ones which contain the unknown b, upon reduction, we obtained one term, hence: $3b - b + 5b = 7b$ with the free terms: $9 - 3 = 6$.

The opposite of a is $-a$. The opposite numbers are reduced by crossing them out with a slash.

summand + summand = sum ,   minuend − subtrahend = difference

The properties of addition :

a) commutativity : $a + b = b + a$, $\forall$ a, b $\in$ **N**   ($\forall$ is the symbol *for all*)
b) associativity: $a + (b + c) = (a + b) + c$, $\forall$ a, b, c $\in$ **N**
c) the neutral element for addition and subtraction is : $a + 0 = a$; $0 + a = a$, $\forall$ a$\in$ **N**.

1.3.2 Multiplication and division

The result of multiplication is called a product.
Multiplicand · multiplier = product.
The numbers which are multiplied are called factors. Let us not confuse the terms from multiplication with the factors from multiplication!

Eg.: $3 \cdot 4 = 12$, 3 is called a multiplicand, 4 is a multiplier and 12 is a product. The numbers 3 and 4 are factors.

The result of division is called a quotient.
Dividend : divisor = quotient.

Zero divided by any natural number different from zero gives us zero.

We observe the following:
$2 \cdot 3 = 3 \cdot 2$ (= 6 in both situations)
$(3 \cdot 2) \cdot 5 = 3 \cdot (2 \cdot 5)$ (= 30 in both situations)
$4(2 + 3) = 4 \cdot 2 + 4 \cdot 3$ (= 20 in both situations)

The properties of natural numbers' multiplication:

a) commutativity : $a \cdot b = b \cdot a, \quad \forall a, b \in \mathbf{N}$
b) associativity : $(a \cdot b) \cdot c = a \cdot (b \cdot c) , \forall a, b, c \in \mathbf{N}$
c) the neutral element is one : $a \cdot 1 = 1 \cdot a = a, \forall a \in \mathbf{N}$
d) the product between any natural number and 0 is 0; $a \cdot 0 = 0, \forall a \in \mathbf{N}$
e) the distributive property of multiplication as opposed to addition and subtraction:
$a(b + c) = ab + ac$
$a(b - c) = ab - ac, \forall a, b, c \in \mathbf{N}.$

Division by zero is not defined.

1.4 The comparison of natural numbers

If **a** is smaller than **b** we write: **a < b**. Some students confuse the sign smaller than (<) with the sign greater than (>). On the natural numbers axis we observe that the more the numbers are to the left, the smaller they are, and the more they are two the right, the larger they are, hence:

| 0 | 1 | 2 | 3 | 4 | 5 | 6 | 7 | 8 | 9 | 10 | 11 | 12 | 13 | 14 | 15 |

                 <   >

Example: $5 > 2$ and $3 < 7$.

For any two natural numbers a and b, there exists one of the following relations:

$a < b$ (a is smaller than b, example: $3 < 5$)
$a = b$ (a is equal to b, example: $3 = 3$)
$a > b$ (a is greater than b, example: $5 > 4$)
$a \leq b$ (a is smaller than or equal to b, example: $5 \leq 5$)
$a \geq b$ (a is greater than or equal to b, example: $12 \geq 9$)

The equality as well as the inequality of natural numbers have the property called **transitivity:**

1) if $a < b$ and $b < c$, then $a < c$, eg: $2 < 3$ and $3 < 5$, then $2 < 5$.
2) if $a \leq b$ and $b \leq c$, then $a \leq c$, eg: $2 \leq 3$ and $3 \leq 5$, then $2 \leq 5$.
3) if $a = b$ and $b = c$, then $a = c$.

This property is frequently used when we have to compare powers and we cannot reduce to the same exponent or the same base.

1.5 Common factor

$$ab + ac = a(b + c) \qquad \text{or} \qquad ab - ac = a(b - c)$$

The common factor is a.

Example: $18a + 45b = 9 \cdot 2a + 9 \cdot 5b = 9(2a + 5b)$. The common factor is 9.

Exercises

1) If $x = 9$ and $a + b = 5$, then $4x + 3a + 3b = ?$
We observe the common factor 3, so:
$4x + 3(a + b) = 4 \cdot 9 + 3 \cdot 5 = 36 + 15 = 51$

2) Calculate 2xa + 3xb + 4a + 6b knowing that 2a + 3b = 13 and x = 6.

For the first two terms we have x as a common factor and for the next two terms we have 2 as a common factor, so:

2xa + 3xb + 4a + 6b = x(2a + 3b) + 2(2a + 3b) = 6 · 13 + 2 · 13 = 78 + 26 = 104.

1.6 The remainder theorem

Dividend = divisor · quotient + remainder, with the property that the remainder < divisor .

$$A = D \cdot Q + R \qquad 0 \leq R < D$$

When we do not have a remainder (reminder = 0), for the division, then we can write: $A = D \cdot Q$

Exercises

The sum of two numbers is 26. If we divide the greater one by the smaller one, we obtain the quotient 7 and the remainder 2. What are the numbers ?

a + b = 26

a : b = 7 and r = 2 so we have a division with a remainder, we will apply the remainder theorem a = b · 7 + 2, 2 < b

If we replace a from a = b · 7 + 2 in a + b = 26, we will obtain:

7b + 2 + b = 26 ⇒ 8b = 26 – 2 ⇒ 8b = 24 ⇒ b = 24 : 8 ⇒ b = 3

and if we replace b in a = 7b + 2, we will have a = 7 · 3 +2 ⇒ a = 21 + 2, so a = 23 and b -3.

1.7 Equations and inequations

The equation is an equality with at least one unknown.

The general form of the equation is:

$$ax + b = 0,$$

a is called a coefficient, **x** is an unknown, root or solution, and **b** is called a free term.

The inequation is an inequality with at least one unknown.

The general form of the inequation is:

$ax + b > 0$, or $ax + b < 0$,

or we can have the signs ≥ (larger or equal to) or ≤ (smaller or equal too).
When we say that the expression is:
- strictly positive, we use the sign > 0
- strictly negative < 0
- positive ≥ 0
- negative ≤ 0 .

We can solve an equation or an inequation using two methods:

1) the inverse operation method. We separate the known elements from the unknowns; when we transpose from one side to the other of the equal or unequal sign, we change the sign.

Example:

a) $x + 8 = 12$, $x = 12 - 8$, we transposed 8 from the left hand side, where it had a +ve sign, to the right hand side, with the –ve sign and we obtained $x = 4$.
b) $2x - 3 = 15$, $2x - 15 + 3$, $2x = 18$, $x = 18 : 2$, $x = 9$.
c) Solve the inequation in the set of natural numbers.
$x < 5 - 2$, $x < 3$, hence $x \in \{0, 1, 2\}$
d) Solve the inequation $3x + 5 > 34 + 2x$, in the set of natural numbers.
$3x - 2x > 34 - 5$, $x > 29$, $x \in \{30, 31, 32, 33,...\}$
e) $4x - 21 \leq 3$, $4x \leq 21 + 3$, $4x \leq 24$, $x \leq 24 : 4$, $x \leq 6$, $x \in \{0, 1, 2, 3, 4, 5, 6\}$, including 6 because we used the sign smaller or equal to, hence it being equal as well, it contains 6.
At d) the sign being only larger than, without the equal, it doesn't contain 29 as well.
f) $3(2x + 5) - 7 \leq 2x + 20 \Rightarrow 6x + 15 - 7 \leq 2x + 20 \Rightarrow 6x - 2x \leq 20 - 8$ (we obtained 8 from $15 - 7$) $\Rightarrow 4x \leq 12 \Rightarrow x \leq 3$, so $x \in \{0, 1, 2, 3\}$.

2) the balance method. The operation that is done on the left hand side of the equal or unequal sign, is done on the right hand side of the equal or unequal sign, which is why we call this the balance method.

If we think about some scales, we will better understand how this method works. For the balance method, we use the sign | and to the right of the sign, we write the operation which is done to the left and right hand sides of the equal sign.

We will work out the same exercises, but this time we will work them out using the balance method.

a) $x + 8 = 12$ | $- 8$, $x = 4$ (8 has been subtracted from the left hand side and the right hand side of the equal sign).

b) $2x - 3 = 15$ | $+ 3$, the following step needs to be thought without being written, but I will write it down so that you understand it better, so: $2x - 3 + 3 = 15 + 3$ (- 3 and + 3 are reduced, because they are the same numbers with opposite signs, so $-3 + 3 = 0$) $\Rightarrow$ $2x = 18$ | $: 2$ (so we divide by 2 on the left hand side and the right hand side of the equal sign) $\Rightarrow$ $x = 9$.

c) Solve the inequation in N $x + 2 < 5$ | -2. We obtain $x < 3$,

$x \in \{0, 1, 2\}$.

d) Solve the inequation in N $3x + 5 > 34 + 2x$ (Usually, for inequations we mostly use the balance method when we have the unknown only on one side of the inequation; however we will work this exercises out using the balance method as well, so you can understand this method well). $3x + 5 > 34 + 2x$ | -2x $\Rightarrow$ (this is thought, that's why I'm writing it in brackets, but in order for it to be clear, I will write it down: $3x - 2x + 5 > 34 + 2x - 2x$) $\Rightarrow x + 5 > 34$ | -5 $\Rightarrow x > 29$, so

$x \in \{30, 31, 32, 33,...\}$.

e) Solve the following inequation in N: $4x - 21 \leq 3$.

$4x - 21 \leq 3$ | +21 , $4x \leq 24$ | :4 , $x \leq 6$,

$x \in \{0, 1, 2, 3, 4, 5, 6\}$.

1.8 Solving problems with the help of equations

When solving problems with the help of equations, we need to respect the following steps:
1) establishing the unknown
2) translating the problem into an equation
3) solving the equation
4) interpreting the result; step 5 - checking, is not compulsory.

Example:

a) Maria has 231 books in her library. If she would give Ioana 31 books, then Ioana would have the same number of books in her library that Maria has in hers. How many books did Ioana initially have?

The steps: 1) $x =$ the number of books Ioana initially has

2) $x + 31 = 231 - 31$

3) $x + 31 = 200$, $x = 200 - 31$, $x = 169$

4) Ioana had 169 books.

b) If we allocate 2 students per bench, 3 students remain unseated. If we allocate 3 students per bench, one bench will be occupied by a student and 3 benches will be free. How many benches and how many students are there in the classroom?

The steps:

1) What am I denoting by x? Since I know how many students I allocate to each bench, if I would know the number of benches, then I could calculate the number of students.

Example: I have 4 benches and I allocate 2 students per bench, so I will have $4 \cdot 2$ students, so 8 students. Everything that I've written until now for step 1 is what I have to think about, so I will note the number of benches with x.

2) I will think about whether or not I'm using the all the benches. Yes, so $x \cdot 2 =$ the no. of students seated on the benches and 3 are standing, so $2x + 3 =$ the no. of students in the classroom.

For the second phase I will think about how many benches will be used: 3 benches are not used, one of them will have one student. So with 3 students, there will be $x - 4$ benches, $3(x - 4)$ represents the number of students who sit 3 per bench and there is one more student who sits alone on a bench, so: $3 (x - 4) + 1 =$ the no. of students in the classroom. Hence, the equation is $3 (x - 4) + 1 = 2x + 3$

3) We multiply the bracket out by applying distributivity and the solution of the equation is: $3x - 3 \cdot 4 + 1 = 2x + 3 \Rightarrow 3x - 12 + 1 = 2x + 3 \Rightarrow 3x - 2x = 12 - 1 + 3 \Rightarrow x = 14$

4) The number of benches is 14 and the number of students is: 2·14 + 3, so 28 + 3, hence 31 students.

c) The difference between two numbers is 32. Dividing the larger number by the smaller one, we get the quotient 3 and the reminder 6. Find the two numbers.

For this problem, we will use the remainder theorem as well as one of the methods for solving an equation. Actually, there will be 2 equations with two unknowns and we will learn later on that it is called a system of equations.

1) we note the larger number with a and the smaller number with b

2) $a - b = 32$

$a = b \cdot 3 + 6$; from the property of the divisor being larger than the remainder, we realize that $b > 6$.

3) If in the first equation we will replace a with the equality from the second equation. In the first equation, we will obtain: $3b + 6 - b = 32$ $\Rightarrow$ $2b = 32 - 6$ (be careful when changing the sign when transposing the terms from one side to the other of the equal sign) so: $2b = 26$ $\Rightarrow$ $b = 26 : 2$ $\Rightarrow$ $b = 13$.

4)   the smaller number is 13 and the larger number is $a - 13 = 32$ (from the first equation), so $a = 32 + 13$, the larger number is 45.

1.9 The order of operations and the use of brackets

The operations are performed in descending order according to the order, so firstly the ones of **order III, which are the powers and roots**, which are studied later on, then the ones of **order II which are the multiplications and divisions** and then the ones of **order I, meaning addition and subtraction.** If we have calculations which are delimited through different types of brackets, we start by calculating the operations within the round brackets, then the square brackets, and finally, the braces.

Example:

a) Solve the equation in the set of natural numbers:
$3 \{6 + 2 [3 (2x - 1) + 5 (11 - 7) - 20] - 8\} = 48$.

If we divide through balance by 3 and at the same time we respect the order of operations and we make the possible calculation from within the round brackets, we will obtain:

$$6 + 2 [3 (2x - 1) + 5 \cdot 4 - 20] - 8 = 16 \qquad | + 8 - 6$$
$$2 [3 (2x - 1) + 20 - 20] = 18 \qquad | : 2$$
$$3 (2x - 1) = 9 \qquad | : 3$$
$$2x - 1 = 3 \qquad | + 1$$
$$2x = 4 \qquad | : 2$$

And the result is : x = 2. We used the balance method because it was easier.

1.10 The divisibility of natural numbers. Divisor, multiple

We say that the natural number a is **divisible** by the natural number b, if there exists a natural number c, so that $a = b \cdot c$.

The number b is called a **divisor** of a, we write a $\square$ b and we read it as "a is divisible by b" or "a is divided by b" or we write b | a and we read "b divides a".

This means that a is divided exactly to b, and hence the divisor in an exact division, is a divisor of the dividend. The number a is a **multiple** of b and c and the numbers b and c are **divisors** of the number a.

Observation: If a $\square$ b   and   a ≠ 0, then there exists a natural number c, so that a = b · c, with c ≠ 0. Hence, a = b · c ≥ b, because the number c is larger or equal to 1 (being different to zero). It follows that a ≥ b.

For a natural number n, we denote with M_n the set of the multiples of n and with D_n the set of the divisors of n. The number of multiples of a number is infinite; the multiples are obtained by multiplying that number with all the natural umbers. The numbers of the divisors of a number is finite.

Example:

1) $M_2 = \{0, 2, 4, 6, 8, 10,...\}$. $D_6 = \{1, 2, 3, 6\}$. $D_7 = \{1, 7\}$.

So if **a = b · c,** then a $\in M_b$ (which is read "a is a multiple of b"), and a $\in M_c$. Or we can write b $\in D_a$ și c $\in D_a$ (which is read "b is the divisor of a" , "c is the divisor of a").

2) 10□2 (we read „10 is divided by 2"), or 2 | 10 (we read "2 divides 10"), so we can write 10 = 2 · 5, 2 is a divisor of 10 and 5 is a divisor of 10 and 10 is a multiple of 2, but also of 5.

3) The number 9 does not divide by 2 (we write 9 ⫶̸ 2) or 2 does not divide 9 (we write 2 ∤ 9), as 9 does not divide exactly by 2 and there is no natural number which when multiplied by 2 gives 9.

4) The divisors of 30 are: 1, 2, 3, 5, 6, 10, 15 and 30. 30 divides exactly to all these numbers. The multiples of the number 30 are: 0, 30, 60, 90 etc.

1.11 Criteria of divisibility

1. A number can be divided by 2 if the last digit is even $\{0, 2, 4, 6, 8\}$. **Example:** 2634, 536, 112.

2. A number can be divided by 3 if the sum of the digits is a number which divides by 3 (meaning a multiple of 3).

Example: 1521 (1 + 5 + 2 + 1 = 9 and 9□3); 342561 (3 + 4 + 2 + 5 + 6 + 1 = 21 and 21□3).

3. A number divides by 4 if the last two digits form a number which divides by 4, example: 33240 , 546008 , 11236 , 1012.

4. A number divides by 5 if the last digit is 5 or 0. **Example:** 340, 345, 1230, 5625 .

5. A number divides by 9 if the sum of the digits is a number which divides by 9 (meaning is a multiple of 9).

Example: 132156 (1 + 3 + 2 + 1 + 5 + 6 = 18 , 18 = M_9).

6. A number divides by 10 if the last digit is 0.
Example: 120, 34320 , 12200.

7. A number divides by 100 if the last two digits are 00.
Example: 200 ,34200 , 10200.

8. A number divides by 11 if the sum of the even order digits subtract the sum of the odd order digits (so every other digit) gives us a number which divides by 11, meaning M_{11} (including zero).

$\overline{abc}$ □11 so a + c - b = 0, so a + c = b

$\overline{abcd}$ □11 so a + c = b + d.

Example: $2\underline{1}4\underline{3}7\underline{3}4\underline{1}2$ (2+4+7+4+2-1-3-3-1 = 11) , $9\underline{1}8\underline{3}6\underline{0}3$ (9+8+6+3-1-3-0 = 22) $\overline{abc}$ □11 $\Rightarrow$ a+c=b, 297 (2+7=9), $\overline{abcd}$ □11 $\Rightarrow$ a+c = b+d, 2783 (2+8=7+3)

9. A number divides by 25 if the last 2 digits are: 00, 25, 50, 75.
Example: 300, 12325,...

The product of two consecutive numbers is divisible by 2, the product of three consecutive numbers is divisible by 3, the product of 4 consecutive numbers is divisible by 4. Hence, the product of n consecutive numbers is divisible by n.

Exercises

1. Find all the numbers of the form $\overline{5x2}$, knowing that $\overline{5x2}$ □3.

We apply the criterion of divisibility with 3, so $5+x+2 \in M_3 \Rightarrow 7+x = 9$ (as it has to be larger than 7) and then every other 3, so that x is a digit. 7+x=12, 7+x=15 and 7+x=18 is not correct because x does not give us a digit. Therefore we stop. X = {2, 5, 8}.

2. Determine all the natural numbers of the form $\overline{54x}$ which divide by 2 and do not divide by 3. We apply the criterion for the divisibility with 2. x $\in$ {0, 2, 4, 6, 8} $\Rightarrow \overline{54x}$ $\in$ {540, 542, 544, 546, 548}

However the numbers don't have to divide by 3 and 5+4+0=9 , 9□3 and 5+4+6=15 which divides by 3 $\Rightarrow$ x = {2, 4, 8} hence the result is: $\overline{54x}$ = {542, 544, 548}.

3. Determine all the numbers of the form $\overline{1x23y}$ □6 .

A number divides by 6 if it divides by 2 and 3, since 2 · 3 = 6 and the numbers 2 and 3 have no common divisors (except for 1). Because it divides by 2, it means that the last digit is even, hence: y $\in$ {0, 2, 4, 6, 8}. $\overline{1x23y}$ □3 if 1+x+2+3+y $\in$ M₃

y = 0 $\Rightarrow$1+x+2+3+0 $\in$ M₃ $\Rightarrow$ 6 + x $\in$ M₃ $\Rightarrow$ x = {0, 3, 6, 9} $\Rightarrow$ $\overline{1x23y}$ $\in$ {10230, 13230, 16230, 19230}.

y = 2 $\Rightarrow$1+x+2+3+2 $\in$ M₃ $\Rightarrow$ 8+x $\in$ {9, 12, 15} $\Rightarrow$ x $\in$ {1, 4, 7} $\Rightarrow$ $\overline{1x23y}$ $\in$ {11232, 14232, 17232}.

y = 4 $\Rightarrow$ 1+x+2+3+4 $\in$ M₃$\Rightarrow$ 10 + x $\in$ {12, 15, 18}; we can't go on with 21 since 21-10 = 11 and 11 is not a digit, so x = {2, 5, 8} $\Rightarrow$ $\overline{1x234}$ $\in$ {12234, 15234, 18234}.

y = 6 $\Rightarrow$ 1+x+2+3+6 $\in$ M₃$\Rightarrow$ 12 + x = {12, 15, 18, 21} $\Rightarrow$x $\in$ {0, 3, 6, 9} $\Rightarrow$ $\overline{1x236}$ = {10236, 13236, 16236, 19236}.

y = 8 $\Rightarrow$ 1+x+2+3+8 $\in$ M₃$\Rightarrow$14 + x $\in$ {15, 18, 21} $\Rightarrow$ x $\in$ {1, 4, 7} $\Rightarrow$ $\overline{1x238}$ $\in$ {11238, 14238, 17238}.

1.12 Prime numbers and composite numbers

The natural numbers which are different from 0 and 1 have **improper divisors** (these are the numbers 1 and itself) and **proper divisors** (these are other numbers besides 1 and itself).

Example: 1 and 6 are the improper divisors of 6 and 2 and 3 are the proper divisors of 6.

The numbers which only have two divisors, meaning 1 and itself are called **prime numbers**, and the ones which have more than one divisor are called **composite numbers** (we will see that they are composed of prime numbers).

One and zero are not prime or composed, they are **neutral**.

One doesn't have another divisor but itself, and zero has a multitude of divisors but it doesn't have itself because the division by zero is not defined.

The first prime number is 2. **2 is the only prime even number.**

Here are a few prime numbers: 2, 3, 5, 7, 11, 13, 17, 19, 23, 29, 31, 37, 41, 43,...

The number 9 is odd, but it is not prime, since 9 = 3 · 3. The number 15 isn't prime, since 15 = 3 · 5.

In order to find the prime numbers, we apply the Greek mathematician Eratosene's (who lived before our Lord Jesus Christ, between the years 275 – 194) method. This method is called "Erathosene's sieve", because it "sieves" the numbers.

1	2	3	4	5	6	7	8	9	10
11	12	13	14	15	16	17	18	19	20
21	22	23	24	25	26	27	28	29	30
31	32	33	34	35	36	37	38	39	40
41	42	43	44	45	46	47	48	49	50
51	52	53	54	55	56	57	58	59	60
61	62	63	64	65	66	67	68	69	70
71	72	73	74	75	76	77	78	79	80
81	82	83	84	85	86	87	88	89	90
91	92	93	94	95	96	97	98	99	100

Eratostene's sieve.

M_2 apart from 2 is eliminated (⬛), M_3 apart from 3, (⬛), M_5 apart from 5, (⬛), M_7 apart from 7 (⬛). We observe that M_{11} are already eliminated ($22 \in M_2$, $33 \in M_3$, $44 \in M_2$, $55 \in M_5$, etc); M_{13} are already eliminated etc. The numbers which are found in the remaining cells of the tables (the white ones) are prime numbers (they are not multiples of 2, or 3, or 5,…)

And now a few example of composite numbers: 4, 6, 8, 9, 10, 12, 14, 15, 16, 18,...

Any composite natural number can be decomposed in a product of prime factors.

Example: 12 = 2 · 2 · 3; 15 = 3 · 5.

Exercise

Find the prime numbers a, b, c, knowing that a + 2b +2c = 32.

We observe that we have 2 as a common factor. So we obtain:

a + 2(b + c) = 32, 2(b+c) is even and 32 is even. Because two added even numbers give us an even number, it follows that a is even as well. Seeing a is even, the only prime number being 2 $\Rightarrow$ a = 2 $\Rightarrow 2 + 2(b+c) = 32 \Rightarrow 2(b+c) = 32 - 2 \Rightarrow 2(b+c) = 30 \Rightarrow b+c = 15$.

Two added odd numbers give an even number and two added even numbers give an even number. So if the sum is an odd number, it means that one of numbers is even and if it is prime as well, it means that it is 2, hence b = 2 $\Rightarrow 2 + c = 15 \Rightarrow c = 15 - 2 \Rightarrow c = 13$.

We obtain: a = 2, b = 2, c = 13.

1.13 Powers with natural exponents of natural numbers

In order to write 4 · 4 · 4 more easily, we write it as 4^3.

a^n means a · a · a · ... · a n times .

The number **a** is called **the base of the power** and the number **n** is called the **exponent of the power**.

Example: $2^3 = 2 \cdot 2 \cdot 2 \Rightarrow 2^3 = 8$. 2 is called the base, and 3 is the exponent.

Raising to the power is an operation of order III and is performed before the operations of order II (division and multiplication) and the ones of order I (addition and subtraction).

1.14 Rules for performing calculations with powers

1. $a^1 = a$ — any number with the exponent 1 is equal to itself; $2^1 = 2, 3^1 = 3, ...$

2. $1^n = 1$ — for any exponent of 1, the result is 1; $1^0 = 1$, $1^3 = 1$, ...

3. $a^0 = 1$ — for any value of a different from 0; $2^0 = 1$, $34^0 = 1$, ...

4. $0^n = 0$ — for any value of n different from 0; $0^3 = 0$, $0^{123} = 0$, ...

5. $a^m \cdot a^n = a^{m+n}$ — the product between the powers with the same base is also a power with the same base and with the exponent equal to the sum of the powers' exponents; $2^3 \cdot 2^2 = 2^5$

6. $a^m : a^n = a^{m-n}$ — the division of the powers with the same base is also a power with the same base and with the exponent equal to the difference between the exponent of the dividend and the divisor; $2^7 : 2^3 = 2^4$

7. $(a^m)^n = a^{m \cdot n}$ — the power of a power is the power of the same number whose exponent is the product of the exponents; $(2^3)^2 = 2^6$

8. $(a \cdot b)^n = a^n \cdot b^n$ — raising a product of numbers to a power is done by raising each factor of the product to that power:

$2^n \cdot 5^n = (2 \cdot 5)^n = 10^n$

Exercises

1. Calculate: $[(2 \cdot 3^2)^3]^4 - (2^4 \cdot 3^8)^3 = (2^3 \cdot 3^6)^4 - 2^{12} \cdot 3^{24} = 2^{12} \cdot 3^{24} - 2^{12} \cdot 3^{24} = 0$

2. Calculate: $\{[2^3 \cdot (2^2 \cdot 2^3)^4]^2 \cdot (2^8)^3 \cdot 2^7\} : 2^{60} = \{[2^3 \cdot (2^5)^4]^2 \cdot 2^{24} \cdot 2^7\} : 2^{60} = \{[2^3 \cdot (2^5)^4]^2 \cdot 2^{24} \cdot 2^7\} : 2^{60} = \{[2^3 \cdot 2^{20}]^2 \cdot 2^{24} \cdot 2^7\} : 2^{60} = \{[2^{23}]^2 \cdot 2^{24} \cdot 2^7\} : 2^{60} = \{2^{46} \cdot 2^{24} \cdot 2^7\} : 2^{60} = \{2^{46+24-7}\} : 2^{60} = 2^{63} : 2^{60} = 2^3 = 8$

3. $a = [(3^2 \cdot 5^3)^2]^3 : 3^2 - (3 \cdot 5^2)^8$. How many zeroes does the number a finish in?

$a = (3^4 \cdot 5^6)^3 : 3^2 - 3^8 \cdot 5^{16}$ $a = (3^{12} \cdot 5^{18}) : 3^2 - 3^8 \cdot 5^{16}$ $a = 3^{10} \cdot 5^{18} - 3^8 \cdot 5^{18}$

$a = 3^8 \cdot 5^{18}(3^2 - 1)$ $a = 3^8 \cdot 5^{18}(9 - 1)$ $a = 3^8 \cdot 5^{18} \cdot 8$ $a = 3^8 \cdot 5^{18} \cdot 2^3$ $a = 3^8 \cdot 5^{15} \cdot 5^3 \cdot 2^3$

$a = 3^8 \cdot 5^{15} \cdot 10^3$

Hence, a finishes in three zeroes.

4. How many zeroes does A finish in, if $A = a \cdot b$, knowing that $a = 2^{24} \cdot 3^{12} \cdot 5^{43} \cdot 7^4$ and $b = 2^{33} \cdot 3^{33} \cdot 5^{22} \cdot 7^{11}$?

$A = 2^{24} \cdot 3^{12} \cdot 5^{43} \cdot 7^4 \cdot 2^{33} \cdot 3^{33} \cdot 5^{22} \cdot 7^{11} \Rightarrow A = 2^{24+33} \cdot 3^{12+33} \cdot 5^{43+22} \cdot 7^{4+11} \Rightarrow A = 2^{57} \cdot 3^{45} \cdot 5^{65} \cdot 7^{15}$

Seeing as 10 is formed from 2 and 5, we can bring both 2 and 5 to the power of 57. $A = 2^{57} \cdot 5^{57} \cdot 5^8 \cdot 3^{45} \cdot 7^{15} \Rightarrow A = (2 \cdot 5)^{57} \cdot 5^8 \cdot 3^{45} \cdot 7^{15} \Rightarrow A = 10^{57} \cdot 5^8 \cdot 3^{45} \cdot 7^{15} \Rightarrow$ A finishes in 57 zeroes.

1.15 Names of orders of magnitude

Value	Usage in Romania
10^3	Thousand
10^6	Million
10^9	Billion
10^{12}	Trillion ($10^3 \cdot 10^9$ = a thousand billions)

1.16 The decomposition of natural numbers in a product of prime factors

For the decomposition of natural numbers in a product of prime factors we apply the divisibility criteria.

Example:

Decompose the numbers: 3600, 140 and 27984.

Because 3600 finishes with two 0s, remembering powers, we can say that the number 3600 divides by $2^2 \cdot 5^2$, which multiplied give 100, so we are left with 36 and we observe that we can apply the divisibility criterion with 2, and when we cannot apply the divisibility with 2, we will progress to 3, then to 5, 7, 11 (the next prime numbers) and so on until we manage to decompose the number into prime numbers.

3600	$2^2 \cdot 5^2$
36	2
18	2
9	3
3	3
1	

140	$2 \cdot 5$
14	2
7	7
1	

27984	2
13992	2
6996	2
3498	2
1749	3
583	11
53	53
1	

Hence we can write:

$3600 = 2^4 \cdot 3^2 \cdot 5^2; \quad 140 = 2^2 \cdot 5 \cdot 7;$

$27984 = 2^4 \cdot 3 \cdot 11 \cdot 53.$

The number of the divisors of a natural number is calculated as follows: the number is decomposed in prime factors and the exponents are increased by one and multiplied. A number decomposed in prime factors of the type $x^a \cdot y^b$ has divisors $(a+1) \cdot (b+1)$.

Example:

The number of divisors of the number 3600 is

$(4+1) \cdot (2+1) \cdot (2+1) = 45$ because $3600 = 2^4 \cdot 3^2 \cdot 5^2$.

A number is a **perfect square** if it can be decomposed in a power with exponent 2 or in a power which has an even number as an exponent.

Example: $4 = 2^2$, $16 = 4^2$ or $16 = (2^2)^2 \implies 16 = 2^4$ therefore a perfect square can be decomposed in a product of two identical numbers, because $a^2 = a \cdot a$.

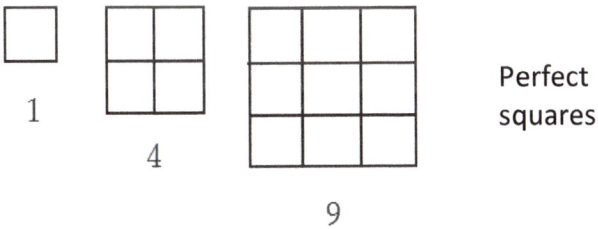

1

4

9

Perfect squares

We observe that $0 \cdot 0 = 0$, $1 \cdot 1 = 1$, $2 \cdot 2 = 4$, $3 \cdot 3 = 9$, $4 \cdot 4 = 16$, $5 \cdot 5 = 25$, $6 \cdot 6 = 36$, $7 \cdot 7 = 49$, $8 \cdot 8 = 64$, $9 \cdot 9 = 81$. Any digit multiplied by itself, gives the last number: 0, 1, 4, 5, 6, 9, therefore **the perfect squares have the last digit 0, 1, 4, 5, 6, 9.** If a number has **the last digit 2, 3, 7 or 8, it is not going to be a perfect square; also a number is not a perfect square if we place it between two consecutive numbers which are perfect squares.**

Example:

1) 124 , we place it between 121 and 144, so $11^2 = 121 < 124 < 144 = 12^2$, therefore 124 is not a perfect square.

2) 12345 is not a perfect square, even if the last digit is 5, because if we apply the divisibility criteria, 12345 is divisible by 5 but is not divisible by 25 or we observe that 1+2+3+4+5 =15 and 15 is divided by 3 but is not divided by 9, hence is not a perfect square.

3) 238 is not a perfect square because the last digit is 8.

4) The number $5n + 3$ is not a perfect square because 5n finishes in 0 if n is even and in 5 if n is odd. If we add 3 to the result we obtain either 3 or 8 for the last digit. Hence, it cannot be a perfect square.

5) The number 9^3 is a perfect square since $9^3 = (3^2)^3 = 3^6$ so it is a perfect square because the exponent is even.

A number is a **perfect cube** if it can be decomposed in a power with **the exponent 3 or a multiple of 3**.

Example: 2^3, 2^6, 4^{12}, etc.

1.17 Common denominator, common multiple

The divisors of 6 are: **1**, **2**, 3, 6. The divisors of 8 are: **1**, **2**, 4, 8. The common divisors of numbers 6 and 8 are 1 and 2. Their greatest common denominator is 2. The least common denominator of two natural numbers is always 1.

The multiples of 2 are: 0, 2, 4, **6**, 8, 10, **12**, ... The multiples of 3 are: 0, 3, **6**, 9, **12**, ... Their least common multiple is 6. The greatest common multiple does not exist.

1) **The greatest common denominator** (abbreviated **g.c.d.** or we write the numbers between round brackets) **is the product of the common prime factors, taken one time at the smallest power.**

2) **The least common multiple** (abbreviated **l.c.m.** or we write the numbers between square brackets) **is the product of the prime, common and uncommon factors, taken one time, at the greatest power.**

3) Two numbers are called **coprime** if they have 1 as their greatest common denominator. The numbers a and b are coprime if g.c.d. (a, b) = 1. This means that they don't have any common denominators larger than 1.

Solved exercises:

1) $(150 ; 504) = ?$ $\qquad 150 = 2 \cdot 3 \cdot 5^2$

$\qquad\qquad\qquad\qquad\quad \underline{504 = 2^3 \cdot 3^2 \cdot 7}$

$\qquad\qquad\qquad\qquad\quad$ g.c.d. $= 2 \cdot 3 \implies (150 ; 504) = 6$

2) $[150 ; 504] = ?$ $\qquad 150 = 2 \cdot 3 \cdot 5^2$

$\qquad\qquad\qquad\qquad\quad \underline{504 = 2^3 \cdot 3^2 \cdot 7}$

$\qquad\qquad\qquad\qquad\quad$ l.c.m. $= 2^3 \cdot 3^2 \cdot 5^2 \cdot 7 \implies [150 ; 504] = 8 \cdot 9 \cdot 25 \cdot 7$

$\quad [150 ; 504] = 12600$

3) Are the numbers 170 and 81 coprime ? $\qquad 170 = 2 \cdot 5 \cdot 17$

$\qquad\qquad\qquad\qquad\qquad\qquad\qquad\qquad\qquad\quad \underline{81 = 3^4}$

$\qquad\qquad\qquad\qquad\qquad\qquad\qquad\qquad\qquad\quad$ g.c.d. $= 1$

Hence, the numbers are coprime.

1.18 The properties of the natural numbers' divisibility relationship

The following properties are very useful for solving exercises:

1. Any natural number is divisible by 1. $a \,\square\, 1 \;\forall\, a \in \mathbf{N}$

Example: $2 \,\square\, 1;\quad 3 \,\square\, 1.$

2. Any natural number is divisible by itself. This property is called reflexivity. $a \,\square\, a \;\forall\, a \in \mathbf{N}$

Example: $3 \,\square\, 3;\quad 8 \,\square\, 8.$

3. The number 0 is divisible by any non-zero natural number.

$0 \,\square\, a \;\forall\, a \in \mathbf{N^*}$

Example: $0 \,\square\, 5$ because $0 = 5 \cdot 0;\quad 0 \,\square\, 12$ because $0 = 12 \cdot 0.$

4. If a is divisible with b and b is divisible by c, then a is divisible by c. This property is called transitivity.

(a ⋮ b și b ⋮ c) ⇒ a ⋮ c

Therefore, if a is divisible with b, then a is divisible by all of b's divisors.

Example: 16 ⋮ 8 and 8 ⋮ 2, so 16 ⋮ 2. The number 16, being divisible by 8, is divisible by all of the divisors that 8 has, hence with 1,2, 4 as well.

5. If a is divisible by b and b is divisible by a, then a = b. This property is called antisymmetry.

(a ⋮ b și b ⋮ a) ⇒ a = b

Explanation: a ⋮ b ⇒ a ≥ b; b ⋮ a ⇒ b ≥ a. From a ≥ b and b ≥ a we get a = b.

6. If a and b are divisible by c, then their sum and their difference is divisible by c.

(a ⋮ c și b ⋮ c) ⇒ a + b ⋮ c

(a ⋮ c și b ⋮ c) ⇒ a - b ⋮ c

Example: 9 and 15 being divisible by 3, their sum, 24, and their difference, 6, are divisible by 3.

7. If a is divisible with b, then the product between a and any number is divisible by b.

a ⋮ b ⇒ a · c ⋮ b

Example: 6 ⋮ 3 ⇒ 6 · 5 ⋮ 3 ⇒ 30 ⋮ 3.

8. If the sum or the difference of two numbers is divisible by a number c and one of the number is divisible by c as well, then the other number is divisible by c as well.

(a + b ⋮ c și a ⋮ c) ⇒ b ⋮ c

(a - b ⋮ c și a ⋮ c) ⇒ b ⋮ c

Example: 14 + b ⋮ 7 and 14 ⋮ 7 ⇒ b ⋮ 7.

9. If a is divisible by the numbers b and c, and these are coprime, then a is divisible by their product as well.

$$(a \, \square \, b \ \text{și} \ a \, \square \, c \ \text{și} \ (b \, ; c) = 1 \ ) \Rightarrow a \, \square \, b \cdot c$$

Example: If a is divisible by 2 and by 3, then a is divisible by 6 as well, because $(2 , 3) = 1$.

10. If $(a \, , b) = d$, then the numbers a and b can be written as $a = d \cdot n, b = d \cdot m, \ \text{cu} \, (n \, , m) = 1$.

Explanation: If $(a \, , b) = d$, then d is a common denominator of numbers a and b, so $a = d \cdot n, b = d \cdot m$ (from the definition of the divisibility relationship). If we wouldn't have $(n \, , m) = 1$, then $(n \, , m) = d_1 \geq 2$. The number d_1, being a divisor of numbers n and m, we would have $n \, \square \, d_1$ and $m \, \square \, d_1$, so $n = d_1 \cdot e, \ m = d_1 \cdot f$. We would have:

$$a = d \cdot n = d \cdot d_1 \cdot e = (d \cdot d_1) \cdot e$$
$$b = d \cdot m = d \cdot d_1 \cdot f = (d \cdot d_1) \cdot f$$

It would result that the numbers a and b have the common denominator $d \cdot d_1 > d$, which would contradict that d is the greatest common denominator.

11. The product between two numbers is equal to the product between the greatest common denominator and the least common multiple.

$$a \cdot b = (a \, , b) \cdot [a \, , b], \ \forall \, a, b \in \mathbf{N}$$

Observation: This formula allows us to rapidly calculate $[a \, ; b]$ knowing $a \cdot b$ and $(a \, ; b)$.

Example: $a = 4, \ b = 6.$ $(4 \, , 6) = 2.$ $[4 \, , 6] = 12.$ We have the equality: $4 \cdot 6 = 2 \cdot 12.$

Solved exercises

1. What are the numbers which have the product 440 and the least common multiple 220?

Solution: Let a and b be the numbers with $a \cdot b = 440$ and $[a, b] = 220$.

The relationship $a \cdot b = (a, b) \cdot [a, b]$ is written as:

$440 = (a, b) \cdot 220 \quad | : 220 \quad \Rightarrow \quad (a, b) = 2 \Rightarrow a = 2x$ and $b = 2y$ with x and y coprime $\Rightarrow a \cdot b = 2x \cdot 2y \Rightarrow 440 = 4xy \Rightarrow xy = 110$.

Bearing in mind the decomposition in prime factors, the number 110 can be written as a product of two factors: $10 \cdot 11$, $2 \cdot 55$, or $5 \cdot 22$. We have the following choices:

$x = 10$ and $y = 11 \quad \Rightarrow \quad a = 20$ and $b = 22$

$x = 2$ and $y = 55 \quad \Rightarrow \quad a = 4$ and $b = 110$

$x = 5$ and $y = 22 \quad \Rightarrow \quad a = 10$ and $b = 44$

2. The sum of two numbers is 50 and the greatest common denominator is 10. What are the numbers?

Solution: $a + b = 50$

$(a, b) = 10 \Rightarrow a = 10x$ and $b = 10y$, with x and y coprime.

The relationship $a + b = 50$ is written: $10x + 10y = 50 \Rightarrow 10(x + y) = 50 \Rightarrow x + y = 5$. We have the following choices:

$x = 1$ and $y = 4$, so $a = 10$ and $b = 40$ (or $x = 4$ and $y = 1$, so $a = 40$ and $b = 10$).

$x = 2$ and $y = 3$, so $a = 20$ and $b = 30$.

3. If the number $7x + 8y$ is divisible by 5, then $2x + 3y$ is divisible by 5 as well.

Solution: $5 \,\square\, 5$, so $5x \,\square\, 5$ and $5y \,\square\, 5$. Through addition, we obtain: $5x + 5y \,\square\, 5$

The numbers $7x + 8y$ and $5x + 5y$ being divisible by 5, their difference is divisible by 5.

$(7x + 8y) - (5x + 5y) \,\square\, 5 \Rightarrow 7x + 8y - 5x - 5y \,\square\, 5 \Rightarrow 2x + 3y \,\square\, 5$.

1.19. Solved exercises

We will now solve a couple of exercises using the theory that we learned until now, we will highlight what is important so you can understand as best as you can.

1. The numbers 1211, 307 and 278 divided by the same number give the remainders 11 ,7 and 8. Find the number they were divided by.

Solution: We observe that in this problem we have divisions with remainders, hence we apply the remainder theorem:
$D = I \cdot C + R$, with the property that $R < I$. Because the greatest remainder is 11, we set the condition $I > 11$. Working through the balance method, we will obtain:

$$1211 = I \cdot C_1 + 11 \quad | \quad -11$$
$$307 = I \cdot C_2 + 7 \quad | \quad -7$$
$$278 = I \cdot C_3 + 8 \quad | \quad -8$$

$$1200 = I \cdot C_1$$
$$300 = I \cdot C_2$$
$$270 = I \cdot C_3$$

We observe that all these numbers have I as a common denominator, therefore we have to calculate g.c.d. (1200, 300, 270). The number I will be a denominator of this g.c.d.

$$1200 = 2^4 \cdot 3 \cdot 5^2$$
$$300 = 2^2 \cdot 3 \cdot 5^2$$
$$270 = 2 \cdot 3^3 \cdot 5$$
$$\overline{\text{g.c.d.} = 2 \cdot 3 \cdot 5 = 30}$$

The number I needs to be searched through the denominators of 30, respecting the condition that $I > 11$.

We obtain $I \in \{15; 30\}$.

For this exercise we used the remainder theorem and g.c.d.

2. Find the smallest natural number which, if divided by 6, 9 and 8 gives the remainder 1.

Solution: We observe here that we have the same remainder as opposed to the previous exercise, but, of course we will still apply the remainder theorem. We presume that the smallest natural number which respects the given conditions is a, therefore:

$$a = 6 \cdot c_1 + 1$$
$$a = 9 \cdot c_2 + 1$$
$$a = 8 \cdot c_3 + 1$$

If the type of quotient is not specified, we will have two cases:

Case I. The quotient is zero. Because 1 : 6, gives the quotient 0 and the remainder, 1 : 9 gives the quotient 0 and the remainder 1, 1 : 8 gives the quotient 0 and the remainder 8, the smallest number will actually be 1.

Case II. The quotient is different from zero. If we move 1 on the left hand side of the equation we will have:

$$a - 1 = 6 \cdot c_1$$
$$a - 1 = 9 \cdot c_2$$
$$a - 1 = 8 \cdot c_3$$

$\Rightarrow a - 1$ is a multiple of 6, 9, 8. Seeing as we want to determine the smallest number with this property, we will take the least common multiple: $a - 1 = [6, 9, 8]$

$6 = 2 \cdot 3$
$9 = 3^2$
$\underline{8 = 2^3}$
l.c.m. $= 2^3 \cdot 3^2$ $\Rightarrow a - 1 = 8 \cdot 9 \Rightarrow a - 1 = 72 \Rightarrow a = 73.$

3. Find the smallest natural number which when divided by 5 gives the remainder 3, when divided by 7 gives the remainder 5, when divided by 9 gives the remainder 7 and when divided by 6 gives the remainder 4.

Solution: We apply the remainder theory again. Denoting the the number we are looking for by a, we will have:

$$a = 5c_1 + 3$$
$$a = 7c_2 + 5$$
$$a = 9c_3 + 7$$
$$a = 6c_4 + 4$$

We cannot put the rest of the left hand side of the equation, because we won't have the same numbers, but we will have a − 3, a − 5 etc. We try to choose a number so that it is a multiple of 5, 7, 9 and 6. We observe that the difference between the divisor and the remainder is the same, 2. Therefore, if through the balance method we add 2, we will obtain:

$a + 2 = 5c_1 + 5$ $a + 2 = 5 (c_1 + 1)$

$a + 2 = 7c_2 + 7$ $\Rightarrow$ $a + 2 = 7 (c_2 + 1)$ $\Rightarrow$ $a + 2 = [5, 7, 9, 6]$

$a + 2 = 9c_3 + 9$ $a + 2 = 9 (c_3 + 1)$

$a + 2 = 6c_4 + 6$ $a + 2 = 6 (c_4 + 1)$

$5 = 5$
$7 = 7$
$9 = 3^2$
$6 = 2 \cdot 3$

l.c.m. $= 2 \cdot 3^2 \cdot 5 \cdot 7 \Rightarrow a + 2 = 2 \cdot 9 \cdot 5 \cdot 7 \Rightarrow a + 2 = 630 \Rightarrow a = 630 - 2$ $\Rightarrow a = 628.$

4. Find all the numbers of the type $\overline{37x7}$ which when divided by 28 will give the remainder 3.

Solution: We have a division with remainder again; after applying that theorem, we will use the decomposition of a natural number in base ten, then the properties of the divisibility relationship:

$\overline{37x7} = 28c_1 + 3 \Rightarrow \overline{37x7} - 3 = 28c_1 \Rightarrow \overline{37x4} = 28c_1 \Rightarrow \overline{37x4} \; \square \; 28.$

Applying the decomposition of a natural number in base ten, we will have: $\overline{37x4} = 3000 + 700 + 10x + 4 = 3704 + 10x \Rightarrow (3704 + 10x) \square \; 28$, and then we apply the remainder theorem to the number 3704 and we obtain $3704 = 28 \cdot 132 + 8 \Rightarrow (28 \cdot 132 + 8 + 10x) \square \; 28.$

Now we need to remember the properties of the divisibility relationships.

$(28 \cdot 132 + 8 + 10x)\ \square\ 28$ and $28 \cdot 132\ \square\ 28$ $\Rightarrow\ (8 + 10x)\ \square\ 28.$

So $10x + 8$ is a multiple of 28, meaning $28 \cdot 0,\ 28 \cdot 1,\ 28 \cdot 2,\ 28 \cdot 3$, etc.

The first choice is $10x + 8 = 0\ \Rightarrow\ 10x = -8$, impossible.
The second choice is $10x + 8 = 28\ \Rightarrow\ 10x = 20\ \Rightarrow\ \mathbf{x = 2.}$
The third choice is $10x + 8 = 56\ \Rightarrow\ 10x = 48$, impossible.
The fourth choice is $10x + 8 = 84\ \Rightarrow\ 10x = 76$, impossible.
The fifth choice is $10x + 8 = 112\ \Rightarrow\ 10x = 104 > 100\ \Rightarrow\ x > 10\ \Rightarrow\ x$ is not a digit.

From here on, we don't obtain digits anymore. This means that the only solution is $x = 2$, so the number we are looking for is $\overline{37x7} = 3727$.

We solved four types of exercises where we applied the remainder theorem. Now we will solve exercises where we will apply the properties of the divisibility relationships.

5. Find two natural numbers different from 0, knowing that their sum is 75, and the g.c.d. is 15.

Solution: If the numbers are denoted a and b, we can mathematically write the hypothesis (the data of the problem) and the conclusion (the requirement).

$a + b = 75$
$\underline{(a, b) = 15}$
$a = ?\ \ b = ?$

We observe that the g.c.d. is 15, so $a = 15x,\ b = 15y,\ (x, y) = 1$.
If we replace $a = 15x$ and $b = 15y$ in the relationship $a + b = 75$, we will obtain:
$15x + 15y = 75\ \Rightarrow\ 15(x + y) = 75\ \ |:15\ \Rightarrow\ x + y = 5\ \Rightarrow$ the solutions are:

$$x = 1\ \Rightarrow\ a = 15 \qquad\qquad x = 2 \qquad a = 30$$
$$y = 4\ \Rightarrow\ b = 60 \qquad\qquad y = 3 \qquad b = 45$$

Observation: The requirement was finding "two natural numbers different from 0", with the mentioned property. Was it necessary to mention that the two natural numbers are different from zero.

If $a = 0$ and $a + b = 75$, then $b = 75$. In this case, the g.c.d. $(a , b) =$ g.c.d. $(0 , 75) = 75 \neq 15$, because $75 \square 75$, $0 \square 75$, and 75 is the greatest with these properties. The condition g.c.d. $(a , b) = 15$ could not have been fulfilled. Hence the condition of the two natural numbers being zero was not necessary. This condition results from the demand that $a + b = 75$ and $(a , b) = 15$.

6. Find two natural numbers knowing that their product is 480 and the l.c.m. is 120.

Solution:

$$a \cdot b = 480$$
$$\underline{[a , b] = 120}$$
$$a = ? \quad b = ?$$

The only formulae from the divisibility relationship which refers to the l.c.m. is:

$a \cdot b = (a , b) \cdot [a , b]$, so :

$480 = (a , b) \cdot 120 \quad | : 120 \Rightarrow 4 = (a , b) \Rightarrow (a , b) = 4 \Rightarrow a = 4x$, $b = 4y$, $(x,y) = 1$.

The relationship $a \cdot b = 480$ is written $4x \cdot 4y = 480 \quad | : 16 \Rightarrow x \cdot y = 30$

$\Rightarrow x = 1 \Rightarrow a = 4 \qquad x = 2 \Rightarrow a = 8 \quad x = 3 \Rightarrow a = 12 \quad x = 5 \Rightarrow a = 20$
$y = 30 \Rightarrow b = 120 \quad y = 15 \Rightarrow b = 60 \quad y = 10 \Rightarrow b = 40 \quad y = 6 \Rightarrow b = 24$

7. There are between 370 and 400 students in the school yard. If we can line them up in rows of 6, 12 and 18 students, how many students are there in the yard ?

Solution: The number of students is a common multiple of 6,12 and 18. We calculate $[6 , 12 , 18]$.

$$6 = 2 \cdot 3$$
$$12 = 2^2 \cdot 3$$
$$\underline{18 = 2 \cdot 3^2}$$
l.c.m. $= 2^2 \cdot 3^2 \Rightarrow [6 , 12 , 18] = 36$.

If in the school yard there are between 370 and 400 students, it follows that there are 36-11 = 396 students.

8. The number $A = 3^8 \cdot 5^7 \cdot 7^9$ is given and the number $B = 2^{11} \cdot 3^7 \cdot 7^{22}$. How many numbers does the number $A \cdot B$ finish with?

Solution:

$A \cdot B = 3^{8+7} \cdot 2^{11} \cdot 5^7 \cdot 7^{9+22} \Rightarrow A \cdot B = 3^{15} \cdot 2^4 \cdot 2^7 \cdot 5^7 \cdot 7^{31} \Rightarrow$

$A \cdot B = 3^{15} \cdot 2^4 \cdot (2 \cdot 5)^7 \cdot 7^{31} \Rightarrow A \cdot B = 3^{15} \cdot 2^4 \cdot 10^7 \cdot 7^{31} \Rightarrow A \cdot B$ finishes in 7 zeroes.

9. Find all the numbers of the form $\overline{4x6y}$ knowing that they are divisible by 6.

Solution:

If a number divides by 6, it means that it divides by 2 and 3 (the divisors of 6, which are coprime $2 \cdot 3 = 6$), so the divisibility criteria with 2 and 3 apply. A number is divisible by 2 if the last digit is 0, 2, 4, 6, 8 and is divisible by 3 if the sum of the digits is divisible by 3.

Therefore, for y = 0, $\overline{4x6y} \,\square\, 3$ if $4 + x + 6 + 0 \in M_3 \Rightarrow 10 + x = $ {12, 15, 18} (x is a digit, hence $x \leq 9$, $10 + x \leq 19$) $\Rightarrow x \in$ {2, 5, 8} $\Rightarrow \overline{4x60} \in$ {4260, 4560, 4860}.

y = 2 $\Rightarrow \overline{4x6y} \,\square\, 3$ if $4 + x + 6 + 2 \in M_3 \Rightarrow x \in$ {0, 3, 6} $\Rightarrow \overline{4x62} \in$ {4062, 4362, 4662}.

y = 4 $\Rightarrow \overline{4x6y} \,\square\, 3$ if $4+x +6+4 \in M_3 \Rightarrow x \in$ {1, 4, 7} $\Rightarrow \overline{4x64} \in$ {4164, 4464, 4764}.

y = 6 $\Rightarrow \overline{4x6y} \,\square\, 3 \Rightarrow$ if $4+x+6+6 \in M_3 \Rightarrow x \in$ {2, 5, 8} , $\overline{4x66} \in$ {4266, 4566, 4866}.

y = 8 $\Rightarrow \overline{4x6y} \,\square\, 3 \Rightarrow 4 + x + 6 + 8 \in M_3 \Rightarrow x \in$ {0, 3, 6, 9}, $\overline{4x68} =$ {4068, 4368, 4668, 4968}.

Concluding: $\overline{4x6y} =$ {4260, 4560, 4860, 4062, 4362, 4662, 4164, 4464, 4764, 4266, 4566, 4866, 4068, 4368, 4668, 4968}.

10. Find the smallest number which divides by 9, 12, 10.

Solution: If this number divides by 9, 10, and 12, it means that it is a multiple of 9, 10 and 12. Because we are asked for the smallest number with this property, we will calculate [9 , 10 , 12].

$$9 = 3^2$$
$$10 = 2 \cdot 5$$
$$\underline{12 = 2^2 \cdot 3}$$

l.c.m. = $2^2 \cdot 3^2 \cdot 5$ ⇒ l.c.m. = 180.   This is the number we were looking for.

11. Find all the numbers of the form $\overline{2x3y} \; \square \; 15$.

Solution: When does a number divide by 15? When it divides by 3 and 5 because $3 \cdot 5 = 15$ and $(3, 5) = 1$.

$\overline{2x3y} \; \square 15$ if it divides by 5 and by 3.

$\overline{2x3y} \; \square 5 \Rightarrow y \in \{0 , 5\}$

$\overline{2x3y} \; \square 3 \Rightarrow 2 + x + 3 + y \in M_3 \Rightarrow 5 + x + y \in M_3$

a) $y = 0 \;\; \overline{2x30} \square 3 \Rightarrow 5 + x \in M_3 \Rightarrow \overline{2x30} \in \{2130, 2430, 2730\}$

b) $y = 5 \;\; \overline{2x35} \square 3 \Rightarrow 5 + x + 5 \in M_3 \Rightarrow \overline{2x35} \in \{2235, 2535, 2835\}$

So $\overline{2x3y} \in \{ 2130, 2430, 2730, 2235, 2535, 2835 \}$

12. How many divisors does the number 936 have ?

Solution: We know that a number which is decomposed in prime factors of the type $x^a \cdot y^b$ has $(a + 1) \cdot (b + 1)$ divisors.

Since $936 = 2^3 \cdot 3^2 \cdot 13$, it has $(3 + 1) \cdot (2 + 1) \cdot (1 + 1)$ divisors, hence it has $4 \cdot 3 \cdot 2$ divisors, so 24 divisors.

13. Write 35^{34} as a sum of two perfect cubes.

Solution:
Let us remember what a perfect cube is. A perfect cube is a number which can be decomposed in the three identical factors or a number which is written in the form a^3.

$35^{34} = 35^{33} \cdot 35 = 35^{33}(27 + 8) = 35^{33}(3^3 + 2^3) = 35^{33} \cdot 3^3 + 35^{33} \cdot 2^3$

$35^{34} = (35^{11} \cdot 3)^3 + (35^{11} \cdot 2)^3$ so we've written it as a sum of two cubes.

14. Write 25^{25} as a two perfect squares sum.

Solution: Knowing that we can write a perfect square in the form a^{2k}, we will think that 25 must have an even exponent, so:
$25^{25} = 25^{24}(16+9) \Rightarrow 2^{25} = 25^{24}(4^2 + 3^2) \Rightarrow 25^{25} = (25^{12} \cdot 4)^2 + (25^{12} \cdot 3)^2$

15. Can you write 65^{31} as a sum of two perfect squares? How about as a sum of two perfect cubes ?

Solution:
$65 = 1 + 64$, $65 = 1^2 + 8^2$, $65^{31} = 65^{30} \cdot 65$, $65^{31} = 65^{15 \cdot 2}(1 + 64)$,
$65^{31} = (65^{15})^2(1^2 + 8^2)$

$65^{31} = (65^{15})^2 + (65^{15} \cdot 8)^2$ sum of two squares.
$65 = 1 + 64$, $65 = 1^3 + 4^3$, $65^{31} = 65^{30} \cdot 65$, $65^{31} = 65^{30}(1^3 + 4^3)$,
$65^{31} = (65^{10})^3(1^3 + 4^3)$
$65^{31} = (65^{10})^3 + (65^{10} \cdot 4)^3$ sum of two cubes

Chapter II. SETS

2.1 Definitions. Operations with sets

A set is made of elements. The set is denoted by uppercase and the elements by lowercase.

Example: A = {a, b} ; B = {3, 7}; C = {a, b, c}; M = {2, 6, 9}.

If a is part of the set M, we write a $\in$ M and read „a belongs to M".

If b is not part of M, we write b $\notin$ M and read „b does not belong to M".

M = {x | P(x)}; x is an element and P(x) is the property through which the set is defined. We read: „M is the set of the elements x with the property P(x)."

Example. M = { x | x $\in$ **N**, x $\leq$ 4} = {0, 1, 2, 3, 4} (M is the set of natural numbers x with the property x $\leq$ 4).

In a set, any element appears only one time. The order of the elements does not matter.

If all the elements of set A belong to set B as well, we say that A is included in B and we write A $\subseteq$ B. If B contains other elements as well, we say that A is strictly included in B and we write A $\subset$ B.

Two sets are equal if they have the same elements. If A $\subseteq$ B and B $\subseteq$ A, then A = B.

Operations with sets

We consider the sets A and B.

The union of sets A and B is the set A $\cup$ B formed from common and uncommon elements, taken one time. Remember that the elements of a set do not repeat!

Example. A = {0, 7, 8}, B = {0, 6, 7, 9}. A $\cup$ B = {0, 7, 8, 6, 9}.

The intersection of sets A and B is the set A $\cap$ B formed only from the common elements.

Example:

A= {1, 5}, B = {5, 8}. A ∩ B ={5}.

A = {0, 1, 2, 3, 4}, B = {0, 2, 4, 6, 8} și C = {0, 1, 2, 4, 5, 7}. A ∩ B ∩ C ={0, 2, 4}.

A = {1, 3} B = {5, 8}.

The two sets do not have any common element. This is why their intersection does not contain any common element. We write: A ∩ B = Φ (the empty set is the one that does not contain any element and is denoted by the Greek letter "phi") A ∩ B = Φ, we say that the sets A and B are **disjoint**.

The difference between sets A and B is the set formed out of the elements which exist in A and do not exist in B.

Example. A = {0, 1, 3, 7}, B = {0, 2, 3}, A \ B = {1, 7}, B \ A = {2}.

If A ⊆ B, we define the **complementary** of A with report to B through A C$_B$A = B \ A.

The symmetrical difference A Δ B (we read "A delta B") is the set formed out of the uncommon elements from both sets.

Example. A = {0, 2, 4}, B = {0, 1, 2, 3}, A Δ B = (A \ B) ∪ (B \ A) = {1, 3, 4}.

The cartesian product of sets A, B represents the set of all the ordered pairs ordonate (a, b) , with a ∈ A , b ∈ B. It is denoted A x B.

Example. A = {0, 1}, B = {1, 2, 3} ⇒ A x B = {(0, 1); (0, 2); (0, 3); (1, 1); (1, 2); (1, 3)}.

The cardinal number of a set A is the number of elements from A and is denoted cardA or |A|.

Example: A = {1, 22, 13, 14, 43}. |A| = 5, because it has 5 elements. |Φ| = 0, because the empty set does not have any elements.

A = { x | x ∈ N , a < x < b } ⇒ CardA = b − a − 1

B = { x | x ∈ N , a ≤ x < b } ⇒ CardB = b − a

C = { x | x ∈ N , a ≤ x ≤ b } ⇒ CardC = b − a + 1

The set of the parts of a set A is $\mathcal{P}(A)$ and contains as elements all of A's subsets, including "the extreme cases", meaning Φ and A.

Example. A = {2, 5}. $\mathcal{P}(A)$ = {Φ, {2}, {5}, {2, 5}}.

The cardinal of set $\mathcal{P}(A)$ is calculated with the $|\mathcal{P}(A)| = 2^{|A|}$. In the previous example, $|\mathcal{P}(A)| = 2^{|A|} = 2^2 = 4$.

2.2 Solved exercises

1) A = {0, 1, 3}, B = {0, 2, 4, 6}. Determine: A∪B , A∩B , A\B, B\A.

A∪B = {0, 1, 3, 2, 4, 6}, A∩B = {0}, A\B= {1, 3}, B\A = {2, 4, 6}.

2) A= {1, 2}, B = {1, 2, 3}. Determine: A∪B, A∩B, A\B, B\A, A x B.

A∪B = {1, 2, 3}, A∩B = {1, 2}, A\B= Φ, B\A= {3}, A x B = { (1,1) , (1,2),(1,3),(2,1),(2,2),(2,3)}

3) A = {1, 3, 5}, B = {1, 2, 3, 4}. We want to find: A∪ B, A∩B, B\A, A Δ B, CardA , Card𝒫(A), 𝒫(A).

A∪B = {1, 2, 3, 4, 5} , A∩B = {1, 3}

B\A = {2, 4} , A Δ B = (A\B)∪(B\A) = {5}∪{2, 4} = {2, 4, 5}, or directly: A Δ B = the uncommon elements, meaning: A Δ B = {2, 4, 5}.

Card A = 3 , Card𝒫(A) = 2^{CardA} = 2^3 = 8, so we have 8 subsets.

𝒫(A) = {Φ, {1} , {3} , {5} , {1, 3} , {1, 5} , {3, 5}, {1, 3, 5}}

4) We know: A∩B = {1, 3, 4} , A\B = { 2 } , B\A = {5, 6, 7}. Determine: A , B , CardA , CardB .

A = {1, 3, 4, 2}, B = {1, 3, 4, 5, 6, 7}, CardA = 4 , CardB = 6

5) If A = { x | x ∈ N and 3 ≤ 2x − 1 < 7 } and B = { x | x ∈ **N** and 4 < 3 x + 1 < 16 }, determine: A∪B , A∩B , A\B, B\A, A Δ B și A x B .

From set A we will solve the inequality as follows:
3 ≤ 2 x − 1 < 7 | +1 ⇒ 4 ≤ 2x< 8 , if we divide by 2, we obtain
2 ≤ x < 4 and as x ∈ N, it follows that A= {2, 3} . From set B we will solve the inequation: 4 < 3x + 1< 16 | − 1 ⇒ 3 < 3x < 15 | :3 ⇒ 1 < x < 5 and x ∈ **N** ⇒ B = {2, 3, 4} so A∪B = {2, 3, 4}, A∩B= {2, 3}, etc.

6) Set A = { x | x ∈ **N**, 2^3 ≤ x ≤ 2^7} is given. Calculate CardA.

CardA = $2^7 - 2^3 + 1$ ⇒ CardA = $2^3(2^4 - 1) + 1$ ⇒
CardA = 8 (16 − 1) + 1 = 8 · 15 + 1 ⇒
We obtain CardA = 121.

7) We consider the sets:

A= { x ∣ x ∈ **N** , 2(3x − 1) + 3(2 − x) ≤ 19 }

B = { x ∣ x ∈ **N***, 5x + 7 < 2(x +3) + x +9 }

Determine: A∪ B, A∩B, A \ B, A Δ B, Card$\mathscr{P}$(A).

In order to find the elements of set A, we need to solve the inequality:

2(3x − 1) + 3(2 − x) ≤ 19, 6x − 2 + 6 − 3x ≤ 19, 3x + 4 ≤ 19, 3x ≤ 19 -4, 3x ≤ 15, x ≤ 5 and as x ∈ **N**, we will have A = {0, 1, 2, 3, 4, 5}.

In order to find set B, we need to solve the inequality:

5x + 7 < 2(x + 3) + x + 9, 5x + 7 < 2x + 6 + x + 9, 5x − 3x < 15 − 7, 2x < 8, x < 4 and as x ∈ **N***, we will have B = {1, 2, 3}.

Hence, A = {0, 1, 2, 3, 4, 5}, B = {1, 2, 3}, A∪ B = {0, 1, 2, 3, 4, 5}, A∩B = {1, 2, 3}, A \ B = {0, 4, 5}, A Δ B = {0, 4, 5}, Card$\mathscr{P}$(A) = 2^{cardA} = 2^6, so set A has 64 subsets.

Chapter III. RATIONAL NUMBERS

3.1 Ordinary fractions. Ratio. Proportions

A part of a whole that has been divided into equal parts is called a **fractional unit.**

Example:

Half of a whole (one part out of two equal parts) is $\frac{1}{2}$ or 1/2, a third is the third part of a whole (one part out of three equal parts) and we write it $\frac{1}{3}$ or 1/3, a quarter is the fourth part of a whole and we write it $\frac{1}{4}$ or 1/4.

Observation: 1/0 (one part out of zero parts) does not make sense.

A pair of natural numbers a and b, with $b \neq 0$, which is written as $\frac{a}{b}$ or a/b (is read "a over b"), is called an ordinary fraction:

a is called a numerator and b is called a denominator. We highlighted **i** from the word denominator because this will help us not confuse the denominator with the numerator, knowing that the one which contains the letter **i**, is below the line of fraction.

Depending on their size with respect to the unit (1), the fractions a/b are divided into three categories:

1) proper fractions (smaller than 1), if a < b; example $\frac{2}{5}$.

2) unit fractions (equal to 1), if a = b; example $\frac{2}{2}$

3) improper fractions (greater than 1), if a > b; example $\frac{5}{2}$

The fractions 1/2 and 2/4 represent the same quantity (1/2 means a half and 2/4 means two quarters, so a half as well). These types of numbers, which are represented by more than one fraction, are called **rational numbers.**

1/2   2/4

We observe that we can write any natural number n as a fraction n = n/1.

The set of rational numbers (positive and negative) is denoted with the letter **Q.** It includes natural numbers (**N** = {0, 1, 2, 3, ...}), whole numbers (**Z** = {0, 1, -1, 2, -2, 3, -3, ...}) and fractional (1/2, 3/5, -8/3 etc.).

Two fractions $\frac{a}{b}$ and $\frac{c}{d}$ are **equivalent** if $\boxed{\frac{a}{b} = \frac{c}{d}}$, (the fundamental property of proportion) so if they represent the same quantity. This can be verified easily as follows: $a \cdot d = b \cdot c$.

Example: 2/3 = 4/6; we have $2 \cdot 6 = 3 \cdot 4$.

The properties of ordinary fractions' equivalence:

1) Reflexivity: $\frac{a}{b} = \frac{a}{b}$

2) Symmetry: If $\frac{a}{b} = \frac{c}{d}$, then $\frac{c}{d} = \frac{a}{b}$ as well.

3) Transitivity: If $\frac{a}{b} = \frac{c}{d}$ and $\frac{c}{d} = \frac{e}{f}$, then $\frac{a}{b} = \frac{e}{f}$.

To amplify the fraction means to multiply both the numerator and the denominator with the same non-zero natural number.

Example:

By amplifying the fraction 5/6 by 2 we obtain 10/12. The amplification is very important because fractions cannot be added together unless they have the same denominator; they are brought to the same denominator through amplification, sometimes, when it can be done, through simplification as well (division of the numerator and the denominator by the same non-zero natural number). In order to bring them to the same denominator, we need to know how to calculate the l.c.m., meaning the least common multiple.

Let's remember how to calculate the l.c.m.: we find the prime factorization of the denominators, and the l.c.m. is the product of the common and uncommon prime factors, taken once, at the highest power.

Every fraction is amplified by those factors from the l.c.m., which we cannot find in the denominator. After multiplication, the denominator has to be equal to the l.c.m.

Example:

$$\frac{3}{4}+\frac{5}{6}=\frac{3\cdot 3}{12}+\frac{2\cdot 5}{12}=\frac{19}{12}$$

l.c. m. (4, 6) = 2² 3 = 12. The first fraction is amplified by 3, and the second one by 2.

If a and b are rational numbers, with b ≠ 0, then $\boxed{\dfrac{a}{b}}$ is called a ratio. A statement of equality between two ratios is called a **proportion**. If a, b, c, d are rational numbers with b ≠ 0 and d ≠ 0, then $\boxed{\dfrac{a}{b}=\dfrac{c}{d}}$ is a proportion.

3.2 Operations with common fractions

3.2.1 Addition and subtraction of fractions

After the fractions are brought to the same denominator, depending on the case, we add or subtract the numerators and copy the common denominator.

$$\frac{a}{b}+\frac{c}{b}-\frac{d}{b}=\frac{a+c-d}{b}$$

Example: $\dfrac{7}{8}+\dfrac{5}{6}+\dfrac{7}{36}-\dfrac{2}{15}+\dfrac{6}{25}=?$

The denominators are decomposed: $8=2^3$, $6=2\cdot 3$, $36=2^2\cdot 3^2$, $15=3\cdot 5$, $25=5^2$. l.c.m. $=2^3\cdot 3^2\cdot 5^2=8\cdot 9\cdot 25 = 1800$

$$\frac{7}{2^3}+\frac{5}{2\cdot 3}+\frac{7}{2^2\cdot 3^2}-\frac{2}{3\cdot 5}+\frac{6}{5^2}=$$

$$= \frac{7 \cdot 3^2 \cdot 5^2}{2^3 \cdot 3^2 \cdot 5^2} + \frac{5 \cdot 2^2 \cdot 3 \cdot 5^2}{2 \cdot 3 \cdot 2^2 \cdot 3 \cdot 5^2} + \frac{7 \cdot 2 \cdot 5^2}{2^2 \cdot 3^2 \cdot 2 \cdot 5^2} - \frac{2 \cdot 2^3 \cdot 3 \cdot 5}{3 \cdot 5 \cdot 2^3 \cdot 3 \cdot 5} + \frac{6 \cdot 2^3 \cdot 3^2}{5^2 \cdot 2^3 \cdot 3^2} =$$

$$= \frac{7 \cdot 9 \cdot 25 + 125 \cdot 4 \cdot 3 + 7 \cdot 2 \cdot 25 - 16 \cdot 3 \cdot 5 + 6 \cdot 8 \cdot 9}{2^3 \cdot 3^2 \cdot 5^2}$$

The fractional numbers can be decomposed in a sum of fractional numbers which have the same denominator,

Example: $\quad \dfrac{9}{5} = \dfrac{2}{5} + \dfrac{3}{5} + \dfrac{4}{5}$

3.2.2 Taking the whole numbers out of the fraction. The introduction of whole numbers into a fraction

The numerator is divided by the denominator. The quotient will represent the whole number, which is written in front of the fraction line, and the rest will be written in the place of the numerator. In the place of the denominator, we write the same denominator.

Example: $\quad \dfrac{17}{5} = 3 + \dfrac{2}{5}$

We write : $\dfrac{17}{5} = 3\dfrac{2}{5}\quad$ and we read „3 whole numbers and two fifths".

In order to introduce the whole numbers into the fraction, we multiply the whole number with the denominator, we add the numerator to the result, and the new result is written in the place of the numerator, while the denominator is the same.

Example: $\quad 3\dfrac{2}{5} = \dfrac{3 \cdot 5 + 2}{5} = \dfrac{17}{5}$

If we have $182\dfrac{5}{6} - 76\dfrac{5}{78}$, it is easier if we don't introduce the whole numbers in the fraction and we bring the fractions to the same denominator.

Here is how it's done: $182\dfrac{5 \cdot 13}{6 \cdot 13} - 76\dfrac{5}{78} = 182\dfrac{65}{78} - 76\dfrac{5}{78} = 106\dfrac{60}{78} = 106\dfrac{10}{13}$

So we subtracted the whole numbers from each other and did the same to the fractions after we brought them to the same denominator.

If the fractions cannot be subtracted, we only introduce a whole number in the fraction, not all the whole numbers, and as such it will be easier for us.

Example:

$$123\frac{5}{6} - 23\frac{73}{78} = 123\frac{5\cdot13}{6\cdot13} - 23\frac{73}{78} = 123\frac{65}{78} - 23\frac{73}{78} = 122\frac{143}{78} - 23\frac{73}{78} = 99\frac{70}{78}$$

Exercises

a) $5\frac{2}{3} + 7\frac{2}{5} =$

b) $21\frac{2}{5} - 3\frac{4}{5} =$

c) $19\frac{3}{4} - 5\frac{5}{6} =$

In order to solve these exercises, it is easier to not introduce the whole numbers in the fraction, and if we cannot do the subtraction, we only introduce one whole number in the fraction. Here it is:

a) We amplify $\frac{2}{3}$ by 5 and $\frac{2}{5}$ by 3. We will obtain :

$$5\frac{10}{15} + 7\frac{6}{15} = 12\frac{16}{15} = 13\frac{1}{15}$$

b) We cannot subtract 4 from 2, so we will only introduce one whole number in the fraction $\frac{2}{5}$, so: $21\frac{2}{5} - 3\frac{4}{5} = 20\frac{7}{5} - 3\frac{4}{5} = 17\frac{3}{5}$.

We have subtracted the whole numbers from the whole numbers and the numerators from each other.

c) We bring $\frac{3}{4}, \frac{5}{6}$ to the common denominator 12 and we obtain:

$19\frac{9}{12} - 5\frac{10}{12}$ and because we cannot subtract 10 from 9, we introduce only one whole number from 19 into the fraction, as it is easier:

$$18\frac{21}{12} - 5\frac{10}{12} = 13\frac{11}{12}.$$

3.2.3 Multiplying the fractions

We multiply the numerators with each other and the denominators with each other:

$$a \cdot \frac{b}{c} = \frac{a \cdot b}{c}$$

$$\frac{a}{b} \cdot \frac{c}{d} = \frac{a \cdot c}{b \cdot d}$$

3.2.4 Simplifying the fractions

Simplifying means dividing both the numerator and the denominator by the same number. For multiplication, we can simplify the numerator from a fraction with the denominator from the second fraction.

Example:

Simplify :

1) $\frac{30}{72}$ We observe that we can simplify by 6 (let's remember the criteria for divisibility!) and we will obtain $\frac{5}{12}$

2) $\frac{15}{12} \cdot \frac{4}{35}$ We observe that we can simplify 15 with 35 by 5, 4 with 12 by 4 and we obtain: $\frac{3}{3} \cdot \frac{1}{7}$. As a result of simplification by 3, we obtain $\frac{1}{7}$.

3) $2\frac{2}{15} \cdot \frac{5}{28}$ First we introduce the whole numbers into the fraction and we obtain $\frac{32}{15} \cdot \frac{5}{28} = \frac{8}{3} \cdot \frac{1}{7}$ because we have simplified 32 with 28 by 4 and 15 with 5 by 5.

The properties of fractional numbers multiplications are the same as for natural numbers.

A fraction which cannot be simplified more is called an **irreducible** fraction. If a fraction can still be simplified, we say that it is **reducible**.

3.2.5 Division of fractions

The inverse of a (a $\in$ N*) is $\frac{1}{a}$ and the inverse of $\frac{b}{c}$ is $\frac{c}{b}$ and $\frac{0}{b}$ does not have an inverse because a fraction which has the denominator 0 does not exist.

$$\frac{a}{b} \cdot \frac{b}{a} = 1, \text{ for a, b} \in \textbf{N*}.$$

In order to divide $\frac{a}{b}$ by $\frac{c}{d}$, we multiply the fraction $\frac{a}{b}$ with the inverse of the fraction $\frac{c}{d}$:

$$\frac{a}{b} : \frac{c}{d} = \frac{a}{b} \cdot \frac{d}{c} = \frac{a \cdot d}{b \cdot c}$$

If the fractional numbers are mixed, we introduce the whole numbers into the fraction, and then we perform the division.

Example: $2\frac{14}{15} : 1\frac{3}{25} = \frac{44}{15} : \frac{28}{25} = \frac{44}{15} \cdot \frac{25}{28} = \frac{11}{3} \cdot \frac{5}{7} = \frac{55}{21} = 2\frac{13}{21}$

$\frac{a}{b} : \frac{c}{d} = \dfrac{\frac{a}{b}}{\frac{c}{d}}$, $\dfrac{\frac{a}{b}}{\frac{c}{d}}$ is called a complex fraction and is equal to: $\frac{a}{b} \cdot \frac{d}{c}$

Example: 1) $4\frac{1}{5} : 1\frac{13}{15} = \frac{21}{5} : \frac{28}{15} = \frac{21}{5} \cdot \frac{15}{28} = \frac{3}{1} \cdot \frac{3}{4} = \frac{9}{4} = 2\frac{1}{4}$

2) $2\frac{1}{3} : 3\frac{8}{9} + \frac{2}{5} = \frac{7}{3} \cdot \frac{9}{35} + \frac{2}{5} = \frac{3}{5} + \frac{2}{5} = \frac{5}{5} = 1$

3.2.6 The natural exponent power of a fractional number

$(\dfrac{a}{b})^0 = 1$ for a ≠ 0 and b ≠ 0. Any number, (different from 0) at the power of 0 is equal to 1.

$\left(\dfrac{a}{b}\right)^n = \dfrac{a^n}{b^n}$ for b ≠ 0   Example. $\left(\dfrac{2}{3}\right)^3 = \dfrac{2^3}{3^3} = \dfrac{8}{27}$

$\left(\dfrac{a}{b}\right)^1 = \dfrac{a}{b}$, $\left(\dfrac{0}{b}\right)^n = 0$ for whatever n ∈ N^* and b ∈ N^*,

$\left(\dfrac{a}{b}\right)^m \cdot \left(\dfrac{a}{b}\right)^n = \left(\dfrac{a}{b}\right)^{m+n}$

$\left(\dfrac{a}{b}\right)^m : \left(\dfrac{a}{b}\right)^n = \left(\dfrac{a}{b}\right)^{m-n}$ with a ≠ 0 , b ≠ 0 m, n∈ **N** and m > n,

$\left[\left(\dfrac{a}{b}\right)^m\right]^n = \left(\dfrac{a}{b}\right)^{m \cdot n}$, $\left(\dfrac{a}{b} \cdot \dfrac{c}{d}\right)^n = \left(\dfrac{a}{b}\right)^n \cdot \left(\dfrac{c}{d}\right)^n$ n ∈ **N**.

Example:

1) $\left(\dfrac{2}{3}\right)^2 \cdot \left(\dfrac{4}{9}\right)^3 = \left(\dfrac{2}{3}\right)^2 \cdot \left[\left(\dfrac{2}{3}\right)^2\right]^3 = \left(\dfrac{2}{3}\right)^2 \cdot \left(\dfrac{2}{3}\right)^6 = \left(\dfrac{2}{3}\right)^8 = \dfrac{2^8}{3^8}$

2) $\left(2\dfrac{1}{2} - \dfrac{2}{3} + \dfrac{1}{6}\right)^3 = \left(\dfrac{5}{2} - \dfrac{2}{3} + \dfrac{1}{6}\right)^3 = \left(\dfrac{15}{6} - \dfrac{4}{6} + \dfrac{1}{6}\right)^3 = \left(\dfrac{12}{6}\right)^3 = (2)^3 = 8$

3) $\left(\dfrac{2}{5}\right)^2 \cdot \left(\dfrac{2}{5}\right)^4 : \left(\dfrac{2}{5}\right)^3 = \left(\dfrac{2}{5}\right)^{6-3} = \left(\dfrac{2}{5}\right)^3 = \dfrac{2^3}{5^3} = \dfrac{8}{125}$

3.3 Comparing the fractional numbers

Out of two fractional numbers with the same denominator, the fraction with the greatest numerator is greater.

Example: $\dfrac{7}{15} > \dfrac{2}{15}$

In order to compare two fractional numbers which don't have the same denominator, we need to bring them to the same denominator.

Example: Compare $\dfrac{23}{15}$ with $\dfrac{25}{16}$,

l.c.m.= $2^4 \cdot 3 \cdot 5 = 16 \cdot 15 = 240$

So: $\dfrac{23 \cdot 16}{240}$ we compare with $\dfrac{25 \cdot 15}{240} \Rightarrow \dfrac{368}{240} < \dfrac{375}{240}$ because

$368 < 375 \Rightarrow \dfrac{23}{15} < \dfrac{25}{16}$

If a, b, c, d are positive numbers we can compare the product of extremes with the medes product: $\dfrac{a}{b} \geq \dfrac{c}{d} \Rightarrow a \cdot d \geq b \cdot c$

3.4 Exercises

1) Determine $x \in N$, for which the following relationships are true:

a) $\dfrac{5}{8} \leq \dfrac{7}{x+3}$ b) $\dfrac{3}{5} \geq \dfrac{x}{20}$ c) $\dfrac{x}{6} \leq \dfrac{3}{2}$ d) $\dfrac{12}{x+2} \geq \dfrac{3}{5}$

Solution:

a) $5(x+3) \leq 7 \cdot 8$

$\Rightarrow 5x + 15 \leq 56 \Rightarrow 5x \leq 56 - 15 \Rightarrow 5x \leq 41 \Rightarrow x \leq \dfrac{41}{5} \Rightarrow x \leq 8\dfrac{1}{5} \Rightarrow x \in \{0,1,2,3,4,5,6,7,8\}$

b) $\dfrac{3}{5} \geq \dfrac{x}{20} \Rightarrow$ after bringing the fractions to the same denominator and after eliminating it because it is the same, we obtain: 12 $\geq x \Rightarrow x \leq 12 \Rightarrow x \in \{\, 0, 1, 2, 3, 4, 5, 6, 7, 8, 9, 10, 11, 12 \,\}$

c) $\dfrac{x}{6} \leq \dfrac{3}{2} \Rightarrow x \leq 9 \Rightarrow x \in \{\, 0, 1, 2, 3, 4, 5, 6, 7, 8, 9 \,\}$

d) $\dfrac{12}{x+2} \geq \dfrac{3}{5} \Rightarrow 12 \cdot 5 \geq 3 \cdot (x+2) \Rightarrow 4 \cdot 5 \geq x + 2 \Rightarrow 20 - 2 \geq x \Rightarrow 18 \geq x$

$\Rightarrow x \leq 18 \Rightarrow x \in \{0, 1, 2, \ldots, 18\}$

3.5 Finding a fraction from a number. Percentages.

In order to find a fraction from a natural number we multiply the fraction by that particular number.

Example: Calculate $\dfrac{3}{4}$ **of** 80. As you can see, I underlined the word **of**; when we will encounter the word **of** we will perform multiplication, hence $\dfrac{3}{4}$ of $80 = \dfrac{3}{4} \cdot 80 = 3 \cdot 20$ (after simplification) $= 60$.

In order to find a fraction from fraction we multiply the two fractions.

Example. How much does $\dfrac{3}{5}$ represent **from** $\dfrac{25}{6}$? As you can see, I have underlined the word **from** in order to point out that we perform multiplication. $\dfrac{3}{5} \cdot \dfrac{25}{6} = \dfrac{5}{2}$ (after simplification and multiplication).

Exercise: Calculate: $\dfrac{2}{3} of 5\dfrac{1}{7}$.

The percentage is expressed as a fraction with the denominator 100. a % is read „a percent" and it will always be out of a number.

Example: How much does 3% **of** 800 represent? As you can see, I have highlighted the word **of** (so we perform multiplication)

$\Rightarrow \dfrac{3}{100} \cdot 800 = 3 \cdot 8 = 24$

If we have a percent of a percent $\Rightarrow a\% \, of \;\; b\;\% \;\; \Rightarrow \dfrac{a}{100} \cdot \dfrac{b}{100}$

which will always be out of a number, hence we will have multiplication.

Example

1) We obtain juice out of 1200 kg of fruits. Knowing that the losses are 7% of the fruit quantity, how many kg of fruits are lost ?

7% of $1200 = \dfrac{7}{100} \cdot 1200 = 7 \cdot 12 = 84$ kg , so we lose 84 kg .

2) Knowing that 1 kg of fruits cost 2 lei, find how many lei we lose if through the assortment of the rotten fruits, we lose 5%. How many kg of fruits are there left after the assortment, if we bought 200 kg ?

5% of $200 = \dfrac{5}{100} \cdot 200 = 5 \cdot 2 = 10$, so we lose 10 kg of fruits, so

$2 \cdot 10 = 20$ lei

After the assortment, we are left with $200 - 10 = 190$ kg fruits.

3) A person has 4000 lei, out of which they spend 12% in order to buy a bike and 4/5 of the remaining money to buy football equipment. How much does a bike cost, how much does the equipment cost and how much money are they left with ?

A bike costs 12% of $4000 = \dfrac{12}{100} \cdot 4000 = 12 \cdot 40 = 480$ lei

The remaining money = $4000 - 480 = 3520$ lei $\Rightarrow \dfrac{4}{5} \cdot 3520 = 4 \cdot 704 =$

2816 lei, so the equipment costs 2816 lei and they are left with
$3520 - 2816 = 704$ lei

4) After a 20% price increase followed by a 10% price decrease, a product ended up costing 2160 lei. What was the initial price and by what percentage has it changed and which way with respect to the initial price?

Step 1. We denote the initial price by x;

Step 2. We put the word problem into the equation: x + 20% x − 10% (x + 20% x) = 2160

Step 3. Solving the equation :

$$\frac{120x}{100} - \frac{10}{100} \cdot \frac{120x}{100} = 2160 \Rightarrow \frac{120x}{100} - \frac{12x}{100} = 2160 \Rightarrow \frac{108x}{100} = 2160$$

$$\Rightarrow x = \frac{2160 \cdot 100}{108} \Rightarrow x = 2000$$

Step 4. Interpreting the result. The initial price was 2000 lei, the last price being 2160 lei, the difference is 160 lei hence: x% 2000 = 160

$$\Rightarrow \frac{x}{100} \cdot 2000 = 1600 \Rightarrow x = \frac{160}{20} \Rightarrow x = 8 \Rightarrow x\% = 8\%$$ so the price has increased by 8% compared to the initial price.

5) An excursionist covers 20 km in three days as follows: the 1st day he walks 10% of the length of the road, then the 2nd day 5/6 of the remaining length, and the 3rd day, the remaining distance. How many km did he cover each day ?

The 1st day he covered $10\% \cdot 20 \Rightarrow \frac{10}{100} \cdot 20 = 2$ km hence the rest is 18km.

The 2nd day he covered $\frac{5}{6} \cdot 18 = 5 \cdot 3 = 15$ km , and the 3rd day the rest:

$$18 - 15 = 3 \text{ km} .$$

3.6 Decimal fractions

Decimal fractions are the ones with a decimal separator (depending on the locale, it is usually a comma or a dot). They are of type **a.b; a** is called **a whole number** and **b a decimal number**.

Example: 89.123456 ; 89 is called a whole number and 123456 is called a decimal number; 1 is called a decimal place, 2 is called a hundredth, 3 is called a thousandth, 4 tenths of thousandths , 5 hundredths thousandths and 6 millionths.

We can write as many zeros as we want to the right of a decimal fraction, after the last number, without changing the fraction.

Example: 23.14 or 23.140 or 23.1400. So, if the last decimals are 0, they can be removed without changing the fraction.

There are two types of decimal fractions:

1) **finite** decimal fractions (with a finite number of decimals)

Example: 2.14 ; 5.764

2) **periodic** decimal fractions, which are of two types as well:

a) **simple periodic**. Example. $2.(14)$ hence all the decimals are between brackets, which means that they repeat, so $2,(14) = 2,14141414...$

b) **mixed periodic**. Example. $2,3(14) = 2,3141414...$

3.6.1 Transforming ordinary fractions into decimal fractions

A fraction whose denominator is a power of 10 is written as a finite decimal fraction by putting a dot at the numerator so that the number of figures on the right side of the dot is equal to the numbers from the denominator.

Example: $\dfrac{1234}{1000}$ = 1.234 so we have 3 decimals, the number of zeros that occur at the denominator. If the denominator decomposes in factors of 2 and 5, it will be amplified in such a way in which 2 and 5 will have the same power so that $2^{n} \cdot 5^{n} = (2 \cdot 5)^{n} = 10^{n}$

Example:

1) $\dfrac{7}{40} = \dfrac{7}{2^{3} \cdot 5}$ we will amplify by 5^{2} in order to obtain 5^{3}

$$\Rightarrow \dfrac{7 \cdot 5^{2}}{2^{3} \cdot 5^{3}} = \dfrac{7 \cdot 25}{10^{3}} = \dfrac{175}{1000} = 0.175$$

2) $\dfrac{123}{50}$ is amplified by 2 and we obtain $\dfrac{246}{100} = 2.46$ hence we amplify in such a way in which we can obtain a power of 10 at the denominator, meaning 10 or 100 or 1000 or 10000 etc.

3) $2\dfrac{3}{250}$ we amplify the fraction by 4 and we obtain $2\dfrac{12}{1000}$ =

2.012. We observe that it is easier if we do not introduce the whole numbers into the fraction.

Transforming the ordinary fraction into a decimal fraction can also be done through division, but if the denominator is only formed from factors of 2 and 5, it is easier to amplify, in such a way in which 2 and 5 have the same exponent.

3.6.2 Transforming finite decimal fractions in ordinary fractions

A finite decimal fraction is transformed in an ordinary fraction as follows:

We write the number suppressing the dot, for the denominator we write a power of 10 with the exponent equal to the number of decimals of the decimal fraction or we can say that we place 1 followed by as many zeros as the number has.

Example:

1) $12,3421 = \dfrac{123421}{10000}$ 2) $113,023 = \dfrac{113023}{1000}$ 3) $1000,02 = \dfrac{100002}{100}$

4) $0,2 = \dfrac{2}{10}$ 5) $0,034 = \dfrac{34}{1000}$ 6) $109,8009 = \dfrac{1098009}{10000}$

3.7 Comparing decimal fractions. Approximations.

In order to compare two decimal numbers, we first compare their whole number parts. If these are not equal, the decimal number with the greatest whole number part is greater. If the whole number parts are equal, we compare de decimal parts after we bring them to the same number of decimals, if they don't have the same number of decimals, we add as many zeroes as are necessary for them to have the same number of digits in the decimal part.

Example:

1) Compare 12.3405 with 12.37. We observe that we have the same whole number part 12. We will compare the decimal part. The tenths digit is the same, we move on to the hundredths, the hundredth from the first number is 4, whereas for the second it is 7, seeing $7 > 4 \Rightarrow$ 12.37 > 12.3405.
Observe that even if we didn't add the zeroes to the end of the number 12.37, we could still compare the numbers.

2) Compare 102.0015 with 102.00157. The whole number part is the same. We will compare the decimal part; the tenth is the same 0, the hundredth is the same 0, the thousandth is the same 1, the tens of thousandths are the same 5 and the hundreds of thousandths from the first number are missing. So, here we have to put 0 and the first number becomes 102.00150. Now it has the same number of decimals and because up to the second last digit they are the same digits, we compare the last digit from the decimals 0 with 7. Seeing as $0 < 7 \Rightarrow$ 102.00150 < 102.00157. The decimal fraction does not change if we put one or more zero at the end of the decimal part. For example 11.23= 11.230 or 11.23 = 11.2300 .

3) Compare 14.89 with 10.92. Because the whole number part of the first number is greater: 14 > 10, we don't need to compare the decimal part as well $\Rightarrow$ 14.89 > 10.92

Approximations. The number 4.56 can be approximated by rounding down, thus giving 4.50, and by rounding up, thus giving 4.60. When a decimal number is rounded, we approximate to the closest value: the last digit which will remain unchanged if the digit that follows ≤ 4 and grows by 1 if the digit that follows is ≥ 5.

Example:
Approximating to the closest decimal:
1) 1.78 will be 1.8 because the decimal 7 increases by 1 because after it we have a digit larger than 4.
2) 12.24 $\approx$ 12.2 because the next digit is 4 and it is < 5

Rounding to hundredths

1) 132.231 ≈ 132.23   2) 11.128 ≈ 11.13 for the first number, we stopped at the hundredth which we had because after it we had a digit smaller than 5 whilst for the second number the hundredth increased by 1 because after it we had a digit larger than 4.

3.8 Operations with decimal fractions

3.8.1 Addition and subtraction

In order to add or subtract, the finite decimal numbers are placed one below the other in such a way in which the dots occupy the same position, then we add or subtract them in the same manner as for natural numbers.

Example:

1) 231.32 +   **2)** 121.014 +   **3)** 0.0029 +
        12.523   8112.13   91.33
       243.843   8233.144   91.3329

4) 23.001 + 192.22 + 2144.5567 = ?   23.001 +
                                                      192.22
                                                     2144.5567
                                                     2359.7777

5) 154.43-38.158 = ?   We observe that the second number has 3 decimals, whereas the first one has two decimals. Because of that, to the first number, we will add a zero at the end (being the last decimal, the number does not change):

        154.430 -
         38.158
        116.272

6) 0.94 − 0.6507 = ?   0.9400 −
                                0.6507
                                0.2893

3.8.2 Multiplication

a) For the multiplication of a decimal number with a power of 10 we move the dot to the right over as many commas as is the exponent of 10.

Example:

$12.43561 \cdot 10^2 = 1243.561$; $101.783 \cdot 100 = 10178.3$
If we multiplied by 100, we move the dot over two digits to the right because we have two zeroes.

$1.24 \cdot 1000 = 1240$ Here we have to move the dot over three decimals to the right, and since we only have two, the third will be zero.

b) Two decimal numbers are multiplied as follows: we multiply the two numbers the same way we multiply natural numbers, disregarding the dot, and then, for the result we add the dot from the right towards the left over the number of digits corresponding to the number of decimals which the two numbers have together.

Example. $26.31 \cdot 15.2 = ?$ $26.31 \cdot$
$\underline{15.2}$
5262
13155
$\underline{2631}$
399.912

3.8.3 Division

a) The division of a decimal number with a power of 10 is done by moving the dot to the left over a number of digits equal to the exponent of 10. Example: $231.56 : 10^2 = 2.3156$; or, if we divide by 1000, because it has 3 zeroes, the dot will pass over 3 digits to the left; Example: $9234.132 : 1000 = 9.234132$; $13.54 : 10000 = 0.001354$.

In the last case, we observe that we need to move the dot over 4 digits because it divides by 10000 and has 4 zeroes.

b) The division of a decimal number to another decimal number is done as follows: we move the dot to the right, for the dividend and the divisor, over as many digits as the number of decimals the divisor has, transforming the divisor into a natural number, fixing the dot to the quotient when we get to it for the dividend.

Example:

1) $2.25 : 0.5 = 22.5 : 5 = 4.5$
2) $11.85 : 1.2 = 118.5 : 12 = 9.875$
3) $1.2 : 0.05 = 120 : 5 = 24$
4) $20.4 : 1.02 = 2040 : 102 = 20$
5) $1.024 : 0.32 \Rightarrow 102.4 : 32 = 3.2$

3.8.4 The power with a natural exponent of a decimal number

$1.2^2 = 1.2 \cdot 1.2 = 1.44$ or $1.2^2 = \left(\dfrac{12}{10}\right)^2 = \dfrac{12^2}{10^2} = \dfrac{144}{100}$ $or\left(\dfrac{6}{5}\right)^2 = \dfrac{36}{25}$ (this is

after simplifying by 2)

3.9 Periodic decimal fractions

Transforming the irreducible ordinary fractions in periodic decimal fractions

1) If the denominator decomposes in different prime numbers than 2 and 5, then, through division, the fraction transforms into a **simple periodic** fraction.

Example. 1) $\dfrac{2}{3} = 0,(6)$ 2) $\dfrac{5}{7} = 0,(7142857)$ 3) $\dfrac{16}{33} = 0,(48)$

We observe that 33 is formed out of the prime factors 3 and 11.

2) If the denominator is decomposed into prime factors and contains 2 and/or 5 as prime factors as well, the fraction transforms through division into a **mixed periodic function**. If the denominator decomposes into prime factors and if there exist 2 and/or 5 as prime factors, the fraction transforms through division into a **mixed periodic fraction**.

Example. 1) $\dfrac{7}{6} = 1.1\,(6)$ 2) $\dfrac{11}{15} = 0.7(3)$ 3) $\dfrac{25}{22} = 1.1(36)$

Transforming periodic decimal fractions into ordinary fractions

1) $1.1(36) = \dfrac{1136-11}{990} = \dfrac{1125}{990} = \dfrac{25}{22}$ for the numerator we write the number disregarding the dot, we subtract everything that is outside the period, and for the denominator we write a number of 9's equal to the number of decimals we have inside the period and a number of 0's equal to the number of the decimals we have outside the period.

2) $27.(478) = \dfrac{27478-27}{999} = \dfrac{27451}{999}$

3) $8.12(5) = \dfrac{8125-812}{900} = \dfrac{7313}{900}$

4) $0.23(123) = \dfrac{23123-23}{99900} = \dfrac{23100}{99900} = \dfrac{231}{999} = \dfrac{77}{333}$ (this is the result which follows as a consequence of simplification)

3.10 The arithmetic mean

In order to calculate the arithmetic mean of a set of numbers, we calculate the sum of the numbers and we divide by the number of the elements of the set. It is denoted by: m_a .

$$m_a = \frac{x_1 + x_2 + x_3 + ... + x_n}{n}$$

Example:

1) Calculate the arithmetic mean of the numbers: 24.1 ; 12.3; 5.2 and 3.2

$$m_a = \frac{24.1+12.3+5.2+3.2}{4} = \frac{44.8}{4} = 11.2$$

2) The arithmetic mean of three numbers is 20, the arithmetic mean of two out of the three numbers is 25. Find the third number.

$$\frac{a+b+c}{3} = 20 \qquad \frac{a+b}{2} = 25 \quad ; \quad c = ?$$

$a + b + c = 60$ and $a + b = 50$. If we replace a+b into the first equality with 50 we will obtain $50 + c = 60$ hence $c = 60 - 50$, $c = 10$.

3) By how much does the arithmetic mean of the numbers 53, 54, 55, 66 change if we add the number 52 ?

$$m_a = \frac{53 + 54 + 55 + 66}{4} \Rightarrow m_a = \frac{228}{4} \Rightarrow m_a = 57$$

$$m_a = \frac{53 + 54 + 55 + 66 + 52}{5} = \frac{280}{5} = 56$$

$\Rightarrow 57 - 56 = 1$, so it decreases by 1.

Chapter IV. ELEMENTS OF GEOMETRY AND UNITS OF MEASUREMENT

4.1 Point. Line. Plane

The point does not have dimensions. The points are denoted by capital letters.

The line is determined by two distinct points. A line contains an infinity of points.

The lines are denoted by lowercase letters (Fig.1) or by two capital letters (Fig. 2) representing two points from that line. The line is not finite (it is infinite).

Fig. 1   Fig. 2

Three points are **non-collinear**, if they are not on the same line (Fig. 3). We cannot say if two points are non-collinear, because any two points determine a line. The points that are on the same line are **collinear**.

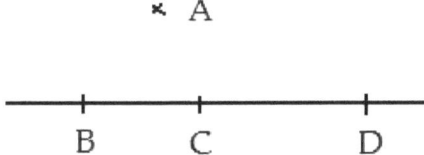

Fig. 3.   B , C and D are collinear. A is non-collinear with B, C, D.

The plane is an extended surface, it is comparable with the surface of a table. Three non-collinear points determine a plane. The plane is denoted by Greek letters: α (alpha), β (beta), γ (gamma) etc.

The lines which intersect each other are called intersecting lines. Parallel lines are the lines situated in the same plane, and which do not have any common points.

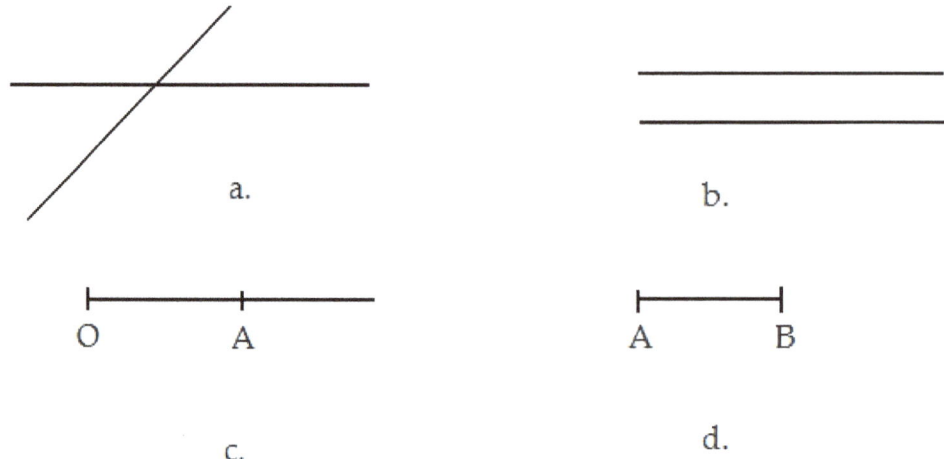

a.   b.

c.   d.

Fig. 4. a. Intersecting lines. b. Parallel lines. c. semi-straight [OA. d. Segment [AB].

[OA is read the **semi-straight line** OA, it is limited by point O and contains point A. Point O is called the origin of **the semi-straight line**.]

The segment AB is written [AB], and the line on which it exists is called **the support line**. A and B are the extremities of the segment. The length of the segment AB is the distance between the points A and B.

4.2 The angle

The angle is the geometric shape formed out of two semi-straight lines which have the same origin. It is read out of a letter if there is no danger of confusion, or out of three letters (the origin is placed in the middle). The angles are:

a) acute (smaller than 90 degrees)

b) right (90 degrees)

c) obtuse (larger than 90 grade and smaller than 180 degrees)

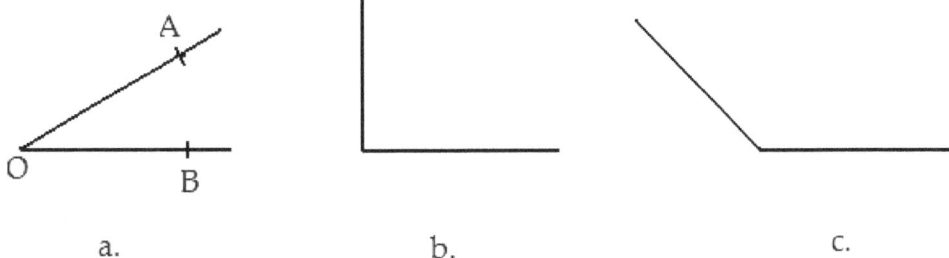

a.   b.   c.

Fig. 5. a. Acute angle. b. Right angle. c. Obtuse angle.

We say that two lines are perpendicular if they form a right angle.

4.3 The triangle

The triangle is the geometric figure made out of three sides and three angles.

The classification of the triangles:

I. by sides:

a) the scalene triangle, has no equal sides. eg. $\triangle$ ABC (Fig. 6)

b) the isosceles triangle, has two equal sides and the angles adjacent to the unequal side are equal, eg. $\triangle$ DEF

c) the equilateral triangle, has all the sides and the angles equal eg. $\triangle$ GHI

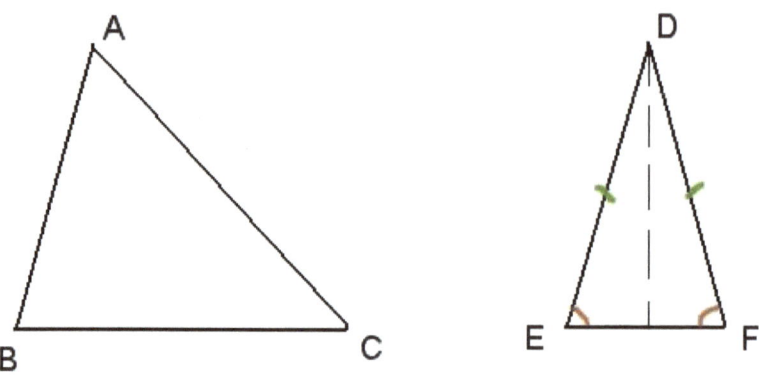

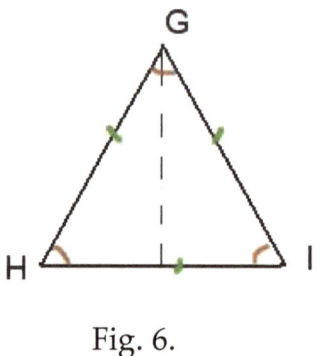

Fig. 6.

II. by angles:

d) the acute triangle, all the angles are acute, eg. Δ JKL (Fig. 7)

e) The right angle triangle, has a right angle, eg. Δ MNP

f) The obtuse triangle, has an obtuse angle, eg. Δ RST

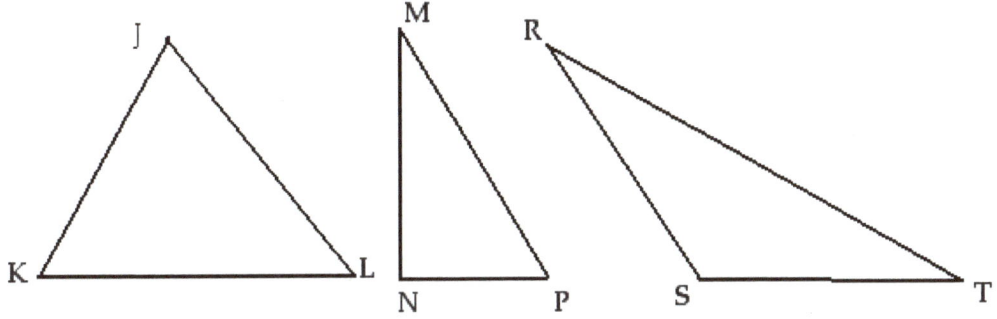

Fig. 7

The sum of the lengths of the sides of a geometric figure is called the **perimeter** of the figure.

The size of the surface of a geometric figure is called the **area**. Two surfaces which have equal areas are called **equivalent**.

The height of a triangle is the **perpendicular** line taken from the peak of the triangle on the opposite side. (Fig. 8).

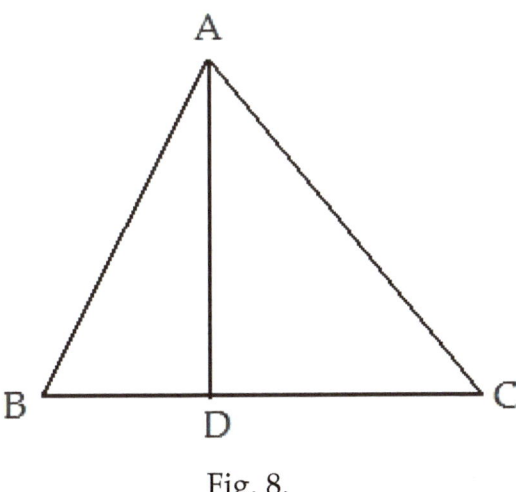

Fig. 8.

The area of the triangle represents the size of the surface of the triangle and we calculate it by multiplying the base with the respective height and we divide the result by 2.

The formula is:

$$A_{\Delta ABC} = \frac{BC \cdot AD}{2}$$

The perimeter is the sum of all the lengths of the sides:

$$P_{\Delta ABC} = AB + AC + BC$$

4.4 Quadrilaterals

The closed plane figure is called a **polygon**. If it has three sides, it is called a **triangle**, if it has four sides it is called a **quadrilateral**.

The elements of a polygon are: the sides of the polygon, the peaks, the angles, the diagonals (the segments with the extremities in non-consecutive peaks).

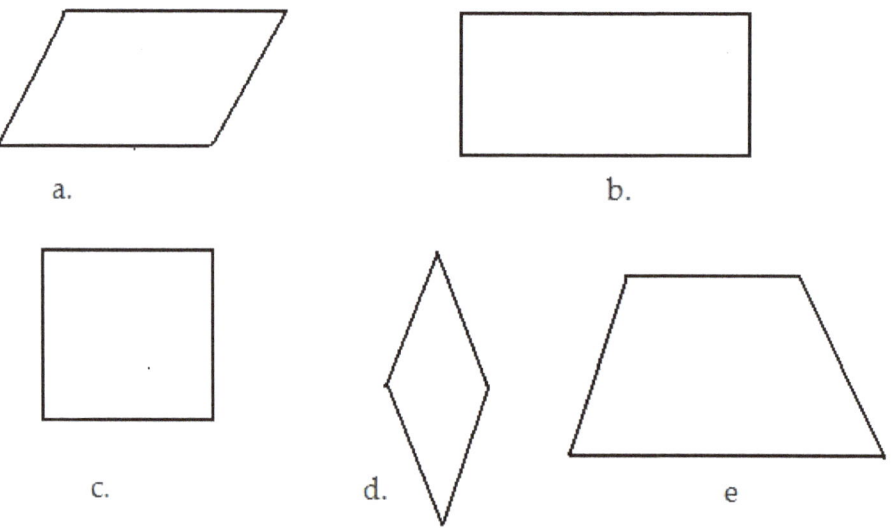

Fig. 9. a. Parallelogram. b. Rectangle. c. Square. d. Rhombus. e. Trapezoid .

a) **The parallelogram** is the quadrilateral with the opposite sides parallel two by two. The opposite sides are congruent as well, meaning they have the same lengths. The largest are called **lengths** and are denoted by L, and the smaller ones are called **widths** and are denoted by l.

The perimeter (P) is the sum of the lengths of the sides and because we have two lengths (L) and two widths (l), $P = 2(L + 1)$.

b) **The rectangle** is the parallelogram which only has right angles (90 degrees). So it has two lengths (L) and two widths (l).

The area is $A = L \cdot l$. The perimeter is $P = 2(L + 1)$

c) **The square** is the rectangle with all the sides of equal lengths; we will denote them with lower-case l. The area of the square is: $A = l^2$ and the perimeter $P = 4l$

d) **The rhombus** is the parallelogram which has all the sides equal.

e) **The trapezoid** is the quadrilateral with two parallel sides (these are considered the bases) and two non-parallel sides.

4.5 The geometric bodies

The cuboid is the geometric figure with dimensions:
The length is denoted by the letter **L**, the width is denoted by **w**, the height is denoted by **h**. The volume is calculated by multiplying these dimensions; the formula is: $V = L \cdot w \cdot h$

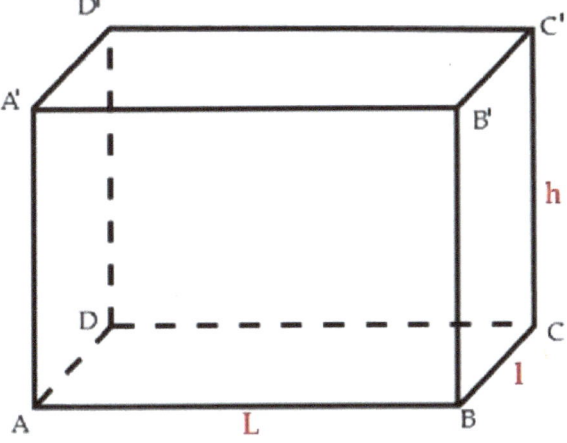

Fig. 10. Cuboid

The cube is the cuboid with equal edges (denoted by l). All the faces are squares. AC and BD are the diagonals of the square ABCD and are perpendicular in point O, which is the middle of the diagonals.

AB = AD = AA′ = the side of the cube= l.

The volume is denoted by V and is calculated by multiplying the length of the edges three times. The formula is: $V = l^3$.

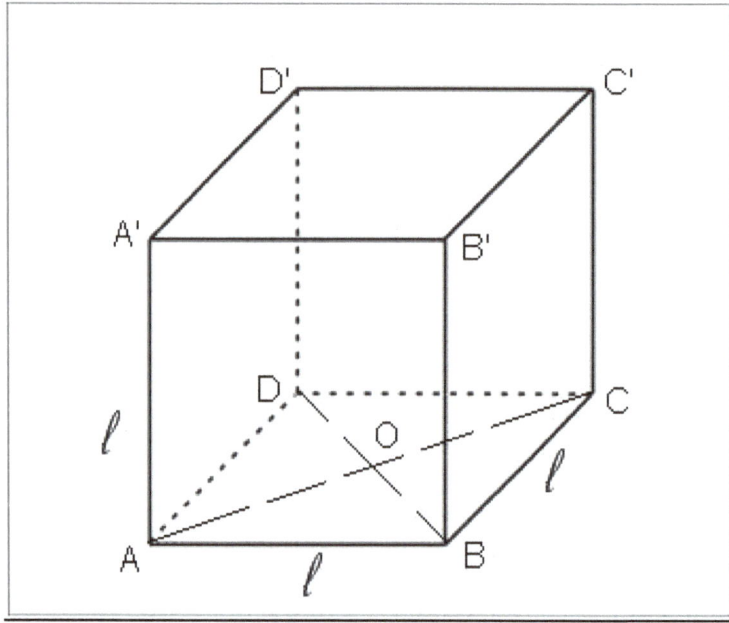

Fig. 11. Cube

4.6. Units of measurement for length

The unit of measurement for length is the **meter (m)**.

On the natural number axis, the greater numbers are noted on the right, and the smaller numbers are noted on the left. This is what we will do in the table below as well.

On the right side, we will write down the multiples of the meter, which are 10 times greater than the meter even if it is right next to the meter, 100 times larger if it is the second multiple after the meter etc.

On the left of the meter, we will write the submultiples which are 10, 100, 1000 times smaller than the meter, if it is next to the meter 10 times smaller, if it is second 100 times, if it is third 1000 times smaller.

The submultiples of the meter				The multiples of the meter		
1 mm	1 cm	1 dm	**1 m**	1 dam	1 hm	1 km
0,001 m	0,01 m	0,1 m	**1 m**	10 m	100 m	1000 m

In order to not mix up the transformation , the students must think as follows:

If they transform **from** greater to smaller, they will multiply by 10 if they transform into the nearby unit, by 100 if the unit is the second, by 1000 if it is the third, (in order to obtain a greater number I need to multiply, which is why I underlined the words **greater** and **multiply;** when I transform I'm thinking that I transform from larger to smaller and I look at the first word which is **greater,** so I **multiply**).

If we transform **from** smaller to greater, we divide (I look at the first word which is **smaller,** so I divide); if it is the first unit with respect to what we're transforming from we divide by 10, if it is the second we divide by 100, if it is the third, we divide by 1000.

Example:

9354.7 m = 9354.7 : 1000 km = 9.3547 km ,

We have transformed from **smaller** to greater, and since the first word is the word **small**, we will divide, because by division we get smaller numbers.

If we think about the first word and if we make the connection between *small* **and** *division* **and between** *division* **and** *great,* **we will not mix up the transformations**.

If the first word is **greater** then we will think that we obtain larger numbers by **multiplying**.

Example:

2.34 m = ……..cm

We transform from **greater** to smaller so we will multiply by 100, next to 1 we write down 2 zeroes because cm are situated on the second place with respect to the m, which is why we will multiply by 100 and so: 2.34m = 2.34 · 100 cm = 234cm.

2546.43cm = ...dam

We transform from **smaller** to greater, so we will perform **division**. Because dam (decameter) is a multiple with respect to the centimeter, it is placed after the decimeter and after the meter, the decameter being the third, we will divide by 10 to the power of three, hence: 2546.43 cm = 2546.43 : 10^3 dam = 2546.43 : 1000 dam = 2.54643 dam .

5786 mm + 28dm + 34.4 dam + 346 cm + 4.8 hm = ...

We will choose to transform either in dm or in dam because they are the units of measurement from the middle. If we transform in dm, we will have: 5786 mm = 57.86 dm ; 34.4 dam = 344 dm ; 346 cm = 34.6 dm ; 4.8 hm = 4800 dm, so:

$$
\begin{array}{r}
57.86 + \\
28 \\
344 \\
34.6 \\
\underline{4800} \\
5264.46 \text{ dm}
\end{array}
$$

A field shaped like a rectangle has the length of 0.25 km and the width of 5000 cm. How much wire do we need to buy if we want to surround the field three times?

Solution:

P = 2 (L + 1),   0.25 km = 250 m and 5000 cm = 50 m , so:

P = 2 (250 + 50) m ;   P = 2 · 300 m ; P = 600 m but we need to surround it 3 times so we will buy 3 · 600 m = 1800 m.

4.7 Units of measurement for the area

The unit of measurement for the area is the square meter (m²).

Submultiples of the square meter			1 m²	The multiples of the square meter		
1 mm²	1 cm²	1 dm²	**1 m²**	1 dam²	1 hm²	1 km²
0,000001 m²	0,0001 m²	0,01 m²	**1 m²**	100 m²	10.000 m²	1.000.000 m²

As we can see, if the unit of measurement is right next to the unit we need to transform, we will divide or multiply by 10^2 (because we have m²); if the unit of measurement will be the second, we will divide or multiply by $(10^2)^2 = 10^4 = 10.000$, and if it will be the third with respect to the unit that we have to transform we will multiply or divide with $(10^3)^2 = 10^6 = 1.000.000$.

When do we divide and when do we multiply ? The same as for the meter. If we transform from **smaller** to greater we will **divide**. As we said before, we will look at the first word, if it is **smaller**, we will divide, if it is **greater**, we will **multiply**.

Example:

1) $238 \, dm^2 = 238 : 10^2 \, m^2 = 2.38 \, m^2$
2) $1435700 \, cm^2 = 1435700 : 10000^2 \, hm^2 = 0.014357 \, hm^2$
3) $3.9824 \, hm^2 = 3.9824 \cdot 1000^2 \, dm^2 = 3982400 \, dm^2$
4) $19.24367 \, dam^2 = 19.24367 \cdot 100^2 \, dm^2 = 192436.7 \, dm^2$

1 ha (hectare) has 10000 m² so: 1 ha = 10000 m.
These units of measurement are used for field surfaces.

Example:

1) How many square meters of tiles do we need to buy for 2 rooms, one shaped like a square with a 4 m side and the other one shaped like a rectangle with the length of 5m and the width of 2.4 m ?

Solution :
The area of the square $= 1^2$ hence : $A_p = 4^2\,m^2$, $A_p = 16\,m^2$ and the area of the rectangle $= 1 \cdot L$, hence $A_d = 5 \cdot 2.4\,m^2$, $A_d = 12\,m^2$, so we need to buy $12\,m^2 + 16\,m^2 = 28\,m^2$ de tiles.

2) Ioana has tomatoes, peppers and cucumbers in her garden. The peppers are planted on a triangle shaped field with the base of 60 dm and the height of 0.04 hm, on each $3\,m^2$ she has 4 pepper stalk. Knowing that on each pepper stalk one can cultivate 2 kg of peppers and that 1 kg of peppers cost 2.5 lei, how much will she earn off of this triangle shaped field? The tomatoes are planted on a rectangular shaped field with the length of 1200 cm and the width of 0.8 dam, on each of the one m^2 she has one tomato vine from which she gathered 3,5 kg tomatoes. Knowing that 1 kg of tomatoes cost 1.5 lei, how much did she earn after she sold the tomatoes? She sold the cucumbers with 2 lei per kg and she gathered from each m^2 3 kg of cucumbers, and they were cultivated on $30\,m^2$. How much did she earn after she sold the cucumbers? What was the total earning?

Solution:
Peppers: Because we cultivate a certain quantity on m^2, we will transform all the dimensions into m and then we will work in m^2, so: The triangle has a base of 60 dm = 6 m , the height of 0.04 hm = 4 m, the area of the triangle is: $A_\Delta = (b \cdot h) : 2$ hence the base times the height and divided by 2, hence $A_\Delta = (6\,m \cdot 4\,m) : 2$, $A_\Delta = 12\,m^2$

If I have 4 pepper stalks on each $3\,m^2$, on the $12\,m^2$, hence $4 \cdot 3$ m^2 she will have $4 \cdot 4$ pepper stalks, hence 16 stalks off of which she will gather $16 \cdot 2$ kg peppers = 32 kg peppers, which cost $32 \cdot 2.5$ lei = 80 lei, hence from the triangle shaped filed she earns 80 lei (for the peppers)

Tomatoes: The area of the rectangle $= L \cdot l$, but we have to transform into m^2, hence L = 1200 cm   L = 1200 : 100 m, L = 12 m and l = 0.8 dam , l = 0.8 $\cdot$ 10. l = 8 m, hence the surface with tomatoes will have

$12 \cdot 8$ m^2 = 96 m^2 off of which she will gather 3.5 kg $\cdot$ 96 = 336 kg tomatoes. If 1 kg of tomatoes cost 1.5 lei so from the 96 m^2 she will earn $1.5 \cdot 336$ = 504 lei.

Cucumbers: From 30 m^2 from where she gathered 3·30 kg of cucumbers, so 90 kg $\Rightarrow$ she earned $2 \cdot 90$ = 180 lei.

So in total she earned: peppers _____ = 80 lei

tomatoes _____ = 504 lei

cucumbers _____ = 180 lei

764 lei

4.8. Units of measurement for volume

The unit of measurement for volume is the meter cubed (m^3).

The following table indicates the units of volume. The transformations are done the same way as the other transformations we presented, the only difference is that we take into account that the exponent is 3 and if we will do transformations into the immediately above or below units (presented in the table) we will **multiply** (if we transform from **greater** to smaller) or we will **divide** (if we transform from **smaller** to greater) with $10^3 = 1000$ (with the exponent 3 because it is a volume).

If we transform a unit of volume into a unit which is on the 2nd place, we will multiply or divide by 100 (2 zeroes because it is number 2 in the table) with the exponent 3, hence 100^3 = 1.000.000, and if the unit we transform in is on the 3rd place, we will multiply or divide by 1000 with exponent 3, hence 1000^3 =1.000.000.000, etc.

The submultiples of the meter cubed			1 m^3	1 dam^3	The multiples of the meter cubed	
1 mm^3	1 cm^3	1 dm^3	**1 m^3**	1 dam^3	1 hm^3	1 km^3
0,000000001 m^3	0,000001 m^3	0,001m^3	**1 m^3**	1000 m^3	1.000.000 m^3	1.000.000.000 m^3

Example :

1) 0.15 dm^3 = 0.00015m^3 . We have transformed from **smaller** to greater, so we divide and the two units are next to each other hence we will **divide** by 10 with the exponent 3 because it is a volume.

2) 0.08 hm^3 = 80000m^3, namely 0.08hm^3 = 0.08 · 100^3m^3, 100 because m is second to hm and we use exponent 3 because it is a volume; we do **multiplication** because we are transforming from **greater** to smaller.

3) 35970612 cm^3 = 35970612: 1000^3m^3 = 0.035970612 m^3. We divided because we are transforming from **smaller** to greater. We divided by 1 followed by 3 zeroes because dam is third to cm, with the exponent 3 because it is a volume.

4) A cuboid has length (L) equal to 350 cm, width (w) equal to 1.2 m and height (h) equal to 20 dm. What is the volume of the cuboid ?
    Because the unit dm is between cm and m we will transform in dm:
L = 350 cm = 350 : 10 dm = 35 dm
w = 1.2 m = 1.2 · 10 dm = 12 dm $\Rightarrow$ V = L · w · h $\Rightarrow V = 35dm \cdot 12dm \cdot 20dm$
h = 20 dm   V = 8400 dm^3 = 8,4m^3

5) A cube of with an edge of 0.004 hm has a volume of …. dm. We traansform 0.004 hm in dm , we multiply because we transform from greater to smaller, we multiply by 1000, we put three zeroes because dm is third to hm, hence the edge of the cube has a length of 4 dm and $V = l^3 = 4^3 dm^3 = 64 dm^3$.

4.9 Units of measurement for capacity

The unit of measurement for capacity is the liter (l). A liter is equal to a cubed decimeter.

The submultiples of the liter				The multiples of the liter		
1 ml	1 cl	1 dl	**1 l**	1 dal	1 hl	1 kl
0,001 l	0,01 l	0,1 l	**1 l**	10 l	100 l	1000 l

1) What is the capacity of a cube with an edge of 200 cm?

The volume of the cube is a^3 if the edge of the cubes is a, but the capacity is measured in liters and seeing as a liter is equal to a dm^3, we need to transform from cm in dm.

200 cm = 20 dm , so $V_{cub} = 20^3 dm^3 \Rightarrow V = 8000 dm^3 \Rightarrow V = 8000$ l

2) What is the capacity of a cuboid if it has the dimensions: L = 0. 008 hm , l = 0.05 dam and h = 600 mm?

$Vcuboid = L \cdot l \cdot h$ but the capacity is measured in liters and as $1 l = 1 dm^3$, we will transform the dimensions in dm:

L = 0,008 · 1000 dm , L = 8 dm

l = 0,05 · 100 dm   , l = 5 dm

h = 600 : 100 dm   , h = 6 dm

$V_{cuboid} = \mathbf{L \cdot l \cdot h}$

$\Rightarrow V_{cuboid} = 8dm \cdot 5dm \cdot 6dm \Rightarrow V_{cuboid} = 240 dm^3 \Rightarrow V = 240$ l

4.10 Units of measurement for mass

The unit of measurement for mass is the gram (g)

The submultiples of the gram				The multiples of the gram		
1 mg	1 cg	1 dg	**1 g**	1 dag	1 hg	1 kg
0.001 g	0.01 g	0.1 g	**1 g**	10 g	100 g	1000 g

Example:

1) 2020 dg + 3020 dag + 28 hg = ... kg

2020 dg = (2020 : 10.000) kg = 0.202 kg

3020 dag = (3020 : 100) kg = 30.2 kg

28 hg = (28 : 10) kg = 2.8 kg

$\Rightarrow$ 0.202 kg + 30.2 kg + 2,8 kg = 33,202 kg

2) Ioana bought 0,15 hg pepper, 1,2 dag bay leaves and 231 cg thyme. How many grams of spices did Ioana buy 0,15 hg piper = 15 g piper

1,2 dag bay leaves = 12 g bay leaves

231 cg thyme = 2,31 g thyme

$\Rightarrow$ 15 g + 12 g + 2,31 g = 29,31 g spices

4.11 Units of measurement for time

The unit of measurement for time is the second (s).

1 minute (min) = 60 s, can also be written as: 1' = 60"

1 hour (hr) = 60 min, or 1 h = 60' = 60 · 60" = 3600"

1 day (d) = 24 hr = 24 · 60' = 1440' = 1440·60" = 86400"

1 week = 7 days

1 year = 365 or 366 days in a leap year (meaning that February has 29 days)

1 decade = 10 year

1 century = 100 year

1 millennium = 1000 year

Was the year 2011 a leap year ?

We need to see whether February had 29 days or 28 days. If February had 29 days, then the year 2011 was a leap year, but since 2011 is not divisible by 4, it means that the year 2011 was not a leap year, because the month of February with 29 days occurs every 4 years; hence if the last two digits of the year are divisible by 4, it means that that year is a leap year.

Example: 2012, 1988, 2004, etc.

www.ingramcontent.com/pod-product-compliance
Lightning Source LLC
Chambersburg PA
CBHW050734180526
45159CB00003B/1225